Safety Systems 2022 Compendium

Published by the Safety-Critical Systems Club 2023.

www.thescsc.org

SCSC-181
ISBN: 9798372158597

Preface

The Safety-Critical Systems Club (SCSC) is the UK's professional network for sharing knowledge about system safety. It brings together engineers and specialists from a range of disciplines and industries working in system safety, academics researching the arena of system safety, providers of the tools and services that are needed to develop the systems, and the regulators who oversee safety.

One of the key outputs of the club intended to promote the sharing of safety engineering best practice is the "Safety Systems" Newsletter. The first issue was published in September 1991 and the Newsletter has been going strong for three decades with Felix Redmill, the first Newsletter Editor, its driving force and champion for the first 25 of those years.

In recent years, new publishing platforms and technologies have been adopted to give the Newsletter a more contemporary look and to make it more widely available in digital form. New editions of the Newsletter are now publicly available immediately for digital download, so all those involved in systems safety can benefit from the content. The annual compendium is however produced for those that would prefer a physical copy of the year's newsletter articles.

The SCSC celebrated its 30th anniversary in 2021, and held its 30th annual Symposium in February 2022, representing a major milestone in the club's history. The February 2022 edition of the newsletter accompanied the Symposium and there were articles reflecting on the achievements of the past 30 years as well as those looking forward to what the next 30 years would bring. A report of the Symposium is provided in the May 2022 edition of the newsletter.

Although Covid-19 has not gone away, 2022 saw it finally loosening its constraining grip on the work of the club, with the first few seminars being held in person during the year. These were blended events with online attendees as well, but they marked an important step in the transition back to the way the club operated prior to the pandemic, with the intention of having a fully in-person symposium in 2023. Reports of all the seminars held over the last year can be found in each of the newsletters published in 2022.

This year saw the launch of the club's "Tech Trips", which are organised visits to places of interest such as museums. The events are open to members and their friends and families and include a dedicated tour guide where appropriate. A report of the first trip to Bletchley Park (home of the World War II codebreakers) and the National Museum of Computing can be found in the May 2022 edition of the newsletter.

The club is always seeking new and engaging content for the newsletter so in the Oct 2022 edition of the newsletter, you can try your hand at the very first Safety Systems crossword!

Paul Hampton
SCSC Newsletter Editor
paul.hampton@scsc.uk

Index of Articles

Volume 30 No. 1

Volume 30 No. 2

Volume 30 No. 3

The Safety-Critical Systems Club Newsletter

Safety Systems

Vol 30 No. 1 - Feb 2022

3 DECADES OF SAFER SYSTEMS

Celebrating 30 years of the SCSC

FUTURE PERFECT?

A new mantra for safety, ethics and sustainability

BACK TO THE FUTURE

Where we're going, we don't need roads...

SAFETY CRITICAL SYSTEMS CLUB 30th SCSC ANNIVERSARY YEAR 30 YEARS OF SAFER SYSTEMS

For everyone working in Systems Safety

SCSC

thescsc.org

SCSC Publication Number: SCSC-171

Contents

WELCOME

FEATURES

REPORTS

GROUPS

EVENTS

The new Safety-Critical Systems eJournal is our latest peer-reviewed publication, containing a blend of industrial papers and academic research results on all aspects of system safety, including the practical aspects – what works and what does not.

Initially, there will be two issues per volume, published in January and July of each year. In addition to the on-line presentation, each Volume of the journal will be made available in printed form each December.

The first issue is out now at https://scsc.uk/journal

If you would like to submit a paper for a future issue, please see "Information For Authors" in the right-hand pane of the journal home page.

Editorial

The SCSC celebrates its 30th Anniversary!

It's remarkable to think that we are now embarking on the 30th volume of the SCSC newsletter, and this month sees the SCSC host its 30th Safety-Critical Systems Symposium (and there's still time to register for the event being held in Bristol and online scsc.uk/e797!)

The club has certainly come a long way in those three decades; it's interesting that many of the standards that are very familiar to us now, like DO-178 and IEC-61508, simply weren't in existence when the club had its inaugural meeting in 1991. Given the unsurpassed attendance at that meeting, there was clearly acknowledgement that there was work to be done, but I suspect there was great uncertainty in what that work was, and what the future had in store. Nevertheless, the club has undoubtedly contributed to a safer world, not only in sharing knowledge of current best practice in system safety, but also in shaping what best practice actually looks like. All of which is certainly a cause for celebration!

The theme for this anniversary edition of the newsletter could be expressed as 'Back to the Future', taking both a backward and forward look at system safety. We will therefore, reflect on the journey the club has taken since its inception along with other historical and more personal accounts, and also look to the future; setting out ideas on how the club could position itself to shape and ensure safer systems for the next 30 years.

Our series of backward-looking articles begins with Tom Anderson and Joan Atkinson, who were key to running the club for most of the club's lifetime. They reflect on the last 30 years of the SCSC and share some history of its formation and entertaining memories of the club's activities and events. This is complemented with an article from Brian Jepson, our webmaster, who describes the evolution of the SCSC website. Stan Price also reflects on his own personal journey in system safety through the years, from those early times when the dependency on computers for safe operations in aircraft was only just being realised.

Our series of forward-looking articles begin with an article from Prof. John McDermid, describing his vision for the role and scope of safety engineering and assurance in a world where wider issues such as ethics and sustainability can no longer be considered separately from traditional safety engineering concerns. John Ridgway presents his vision of the future of Human Factors in his article, and we conclude with a more speculative look at what the future holds as predicted by some of the SCSC Steering Group members.

We also have reports from events held last year on the quantification of risk and the safe use of multicore, and Zoe Garstang provides an update from the Safety Futures Initiative – helping develop the engineers that will be making systems safer over the next three decades.

The future is, perhaps, less murky than it was all those years ago; this gives us more empowerment to shape it, but it also means we are troubled by what we see ahead, especially in relation to wider issues such as climate change. We have come a long way but there is still a long way to go, and there certainly is still much work to be done.

Paul Hampton
SCSC Newsletter Editor
paul.hampton@scsc.uk

In Brief

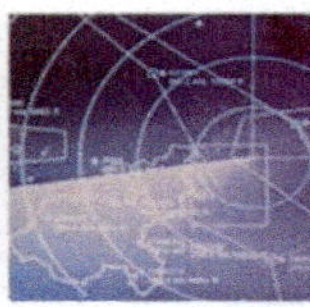

Covid-19: Researcher blows the whistle on data integrity issues in Pfizer's vaccine trial

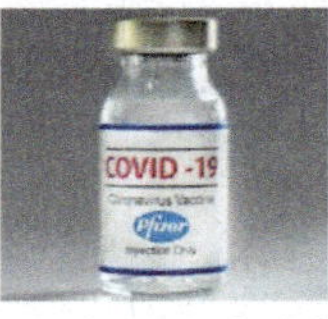

Revelations of poor practices at a contract research company helping to carry out Pfizer's pivotal Covid-19 vaccine trial raise questions about data integrity and regulatory oversight. *bmj.com*

Drone Helps Save Cardiac Arrest Patient in Sweden

For the first time in medical history, a drone has played a crucial part in saving a life during a sudden cardiac arrest. The world unique achievement took place in Sweden when an Everdrone autonomous drone delivered a defibrillator that helped save the life of a 71-year-old man. *uasvision.com*

Smart motorway rollout suspended amid safety concerns

The rollout of smart motorways has been suspended by the government until at least 2025 in response to safety concerns from MPs and motoring groups. *theguardian.com*

Collision between passenger trains at Salisbury Tunnel Junction

Preliminary findings from the investigation of the collision involving two westbound passenger trains at Salisbury's Fisherton Tunnel in October 2021, concludes that the failure of one train to stop at a red light was almost certainly a result of low adhesion between the train's wheels and the rails. *gov.uk*

Tesla driver charged with vehicular manslaughter over fatal Autopilot crash

California prosecutors have filed two counts of vehicular manslaughter against the driver of a Tesla on Autopilot who ran a red light, slammed into another car and killed two people in 2019.

The defendant appears to be the first person to be charged with a felony in the United States for a fatal crash involving a motorist who was using a partially automated driving system. *theguardian.com*

Safe, Ethical & Sustainable: A Mantra for All Seasons?

John McDermid provides some guiding principles on how to achieve and manage the safety of complex systems whose failure causes and consequences go beyond the concerns of traditional safety engineering. He sketches some new models for safety engineering and proposes the adoption of the mantra "safe, ethical and sustainable" to not only focus the attention of the community on the key issues, but also to influence politicians and policies.

There are many principles and "laws" relevant to systems and safety engineering. At a time when the community is grappling with systems of unprecedented complexity I am reminded of Mencken's Law:

"For every complex problem, there is a solution that is clear, simple, and wrong."

There is also the famous [1]:

"All models are wrong, some are useful."

And two further (related) quotes, the first from Humpty Dumpty:

"When I use a word, it means just what I choose it to mean — neither more nor less."

"Words in their primary or immediate signification stand for nothing but the ideas in the mind of him that uses them."

The latter, less well-known perhaps, is from John Locke [2] and may have been the source of Lewis Carroll's ideas for Humpty Dumpty's scornful remark to Alice [3].

Why do I quote these? I want to suggest some new models for how we think about complex problems, with a particular emphasis on safety. These models will be wrong in some facet or some situation, but they will be useful if they help focus thought in a constructive way. And all I have is words. My aim is to convey the "ideas in the mind ..." – but perhaps if they are repeated often enough (like a mantra) they will start to have significance in other minds too. But I start with the usage of a word which I sort of hate.

Elegance in Systems Engineering

Systems engineering has long been presented as a multi-disciplinary, or trans-disciplinary approach to solving complex, multi-faceted problems. To me 'elegance' means being 'graceful or stylish' but the systems engineering community has hijacked it to mean [4]:

- Efficacy – how well does it achieve the desired outcomes?
- Efficiency – how economical is it in use of resources both to develop and to operate it?
- Robustness – how well does the system perform in unanticipated circumstances?
- Minimising unintended consequences – how well does the system do in reducing unwanted and unanticipated consequences?

The aim – akin to mine here – is to give some guiding principles to help achieve focus on the key issues when trying to solve complex problems. I accept that some systems can be elegant, but I think the complexities of modern systems, the difficulties of dealing with brownfield systems design, the interconnectedness and interdependencies of systems, and the fact that efficacious and robust solutions can be downright ugly, makes me think this is a somewhat naïve characterisation. The four principles seem sensible, but I sort of hate the label 'elegant'. Overall, I view this as a good example as it indicates the kind of models I want to create – guiding principles that are useful, although wrong in some sense.

Complex Systems

I have used the term 'complex systems' several times. What do I mean by this? It is common to make a distinction between complicated and complex systems, saying that complex systems exhibit 'emergent properties', that is, properties of the whole that are not simply properties of the individual system components (and their inter-relationships). Some argue that this definition is unscientific – but science doesn't have the answers to everything – moreover, according to Aristotle [5]:

> *"In the case of all things which have several parts and in which the totality is not, as it were, a mere heap, but the whole is something besides the parts."*

I hope I may be excused for relying on this unscientific definition, since the implied uncertainties of 'emergence' indicate one of the key challenges we need to face in safety engineering. We can't assume that the system development will produce results that safely manage all these uncertainties from 'day one', thus we also need to consider the overall governance of the systems in deployment, to identify unintended consequences, to learn from them and provide feedback to improve the system. To me this is another reason that the notion of 'elegance' in systems is somewhat naïve – it implicitly assumes that the system can be designed to achieve its intended use 'full stop'. The reality is, however, that we need to produce systems that are 'good enough' (safe enough) that we can deploy them and then manage them to achieve acceptable safety through life, recognising that what is acceptable may change over time, as technology or society's expectations evolve.

A study on Safety of Complex Systems for the Royal Academy of Engineering [6] introduced a model of how complexity, as opposed to mechanistic failures, contribute to systemic failures, see the figure. It shows the role of both design-time and operation-time controls for reducing the likelihood of systemic failures, or for mitigating the consequences. The exacerbating factors are those issues that can have adverse impacts, particularly on the controls.

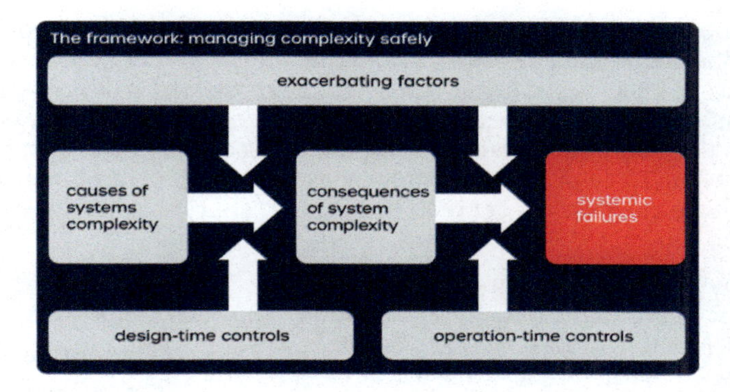

The study also introduced a three-layer model for governance of such systems spanning:

- Governance – cross-jurisdictional incentives and requirements for organisations to adhere to best practice through regulations, standards, soft law, etc.
- Management – risk control and trade-offs in an organisation, management of supply chain dynamics and the sustainment of long-term knowledge
- Task & Technical – the behaviour of the technological elements of the system, the users and other stakeholders, in their context of use

The task & technical level is the traditional province of safety engineering (and it seems the main focus of the work on elegance in systems engineering), but it is insufficient to address and manage the problems of complex systems. The report presents some examples of using the framework, showing the issues in the governance, management and task & technical levels, for each of the elements in the model – causes of complexity, exacerbating factors, etc. For example, in the case of the Uber Tempe accident [7] we can see:

- Exacerbating factor (management layer) – casualisation of labour, using untrained safety drivers (gig economy)
- Causes (task & technical) – mentally unstimulating but critical tasks of system supervision

This work provides two key aspects of the broader models – ways of thinking about systemic failures arising from complexity, and the need to consider causes and controls at the levels of management and governance, as well as those within the more traditional scope of safety engineering. It also highlights the need to consider safety management through life; of course, this is not new, but treating safety management as a continuum rather than having a discontinuity as the system enters service is at least a new emphasis.

Benefits and Harms

Safety mainly focuses on harms to people, i.e. death and injury, and ways of reducing the attendant risk. Of course, benefits are considered, albeit in a simplistic, or limited way. For example, a cost benefit analysis can be used to support a claim that a risk has been reduced As Low As Reasonably Practicable (ALARP). This however is a narrow application focused on risk reduction and doesn't consider the wider societal advantages or benefits of having the system in the first place.

Thus, the proposed shift in our models of safety is to consider benefits as well as harms in all aspects of the design and analysis of systems. In terms of models, this requires definition of a trade-space of benefits and harms that should be considered in designing and analysing systems. This should be much broader than the conventional focus of safety engineering on harm to individuals and should include consideration of society and the environment. A challenge is the need to trade-off between incommensurable factors; this is not easily resolved but doing pairwise comparisons between designs that are close in the trade-space is one possible tactic.

Individual, Societal and Environmental

Although broadening models, it is still desirable to take a human-centric approach, and thus I use well-being as an 'umbrella' [8]. In this content, it is possible to divide benefits and harms into three categories: individual, societal and environmental, with examples of potential benefits and harms shown in the table below. These are intended to be illustrative, and the factors to be taken into consideration would need to be identified for a specific system.

Although broadening [safety] models, it is still desirable to take a human-centric approach

	Benefits	**Harms**
Individual	Personal autonomy Health	Physical injury Mental illness
Societal	Safe working environment Equitable access to resources	Social exclusion Inequitable risk distribution
Environmental	Biological diversity Clean water	Warming of the atmosphere Plastic build-up in the oceans

Many of these benefits and harms have an ethical dimension. For example, if introducing autonomous vehicles net reduces the level of fatal accidents on the roads but does not give any decrease in the number of fatalities for cyclists, then this would be a case of inequitable risk distribution. To be fair to the systems' engineers, this could be seen as an interpretation of what they meant by 'efficacy' and 'minimising unintended consequences'.

Many of these benefits and harms have an ethical dimension

The shift in terms of models supporting safety engineering should be clear – but, maybe now, beginning to seem like 'scope creep' on a global scale!

Toujours L'Attaque Surface

A large proportion of modern engineered systems utilise computer-based control, and many have high degrees of interconnectivity. Communication between systems and between systems and the infrastructure, can help in terms of efficacy and efficiency, but it opens up possibilities of cyber-attacks. Napoleon was famous for saying 'toujours l'attaque' (always attack).

I am advised by people who run industrial and public infrastructure that this idea is alive and well in the "hacker community", and that their systems are under continuous attack. Highly connected systems present a large attack surface, and poor security controls in one part of the overall system may enable access to more critical parts.

As an illustration, the figure shows a teenage boy and the TV remote control that he modified and then used to move points under a tram in Lodz, Poland, causing a derailment.

A further example is the 2001 attack on the Maroochydore sewage plant in Queensland, Australia which released about 1 million litres of sewage [9]. These incidents show the need to consider the interaction of security and safety and there are now many proposals for combined approaches to security and safety analysis, including some focused on early-stage design [10].

Perhaps controversially, I am of the view that the *risks* from cybersecurity are relatively low. The likelihood of attacks on many systems is high, but the probability of success is quite low and the more complex the system, the less likely the attacks are to succeed; in the Maroochydore case, the attacker was a disgruntled former employee, i.e. had inside knowledge. Thus, the models need to include cyber-security, to address means to reduce the potential attack surface and to provide effective security controls. This will reflect legitimate societal concern, but the focus should be on security in the role of its contribution to safety [10].

Safe, Ethical and Sustainable

If we were to draw together the models I have hinted at above, and also provide the supporting detail, e.g. in terms of analysis methods, then it would be clear that they are very complex; indeed, there are over 100 elements in the Safety of Complex Systems framework alone and the other models are multi-dimensional too.

It is not easy to produce a good summary – or mantra – but I propose "safe, ethical and sustainable". Safe – the primary focus continues to be on individual health and safety. Ethical – as new systems exhibit increasing levels of autonomy, moving decision-making from humans to machines, there are many issues including the potential for unfair distribution of risk or, unjustifiably, holding someone liable for outcomes which are beyond their control. Sustainable – due to the importance of sustainability in itself, and the human and societal effects of environmental damage; for example, global warming is already a major source of individual harm [8]. This aligns with the focus on individual, societal and environmental benefits and harms.

It is not easy to produce a good summary – or mantra – but I propose "safe, ethical and sustainable"

Perhaps it is better to view this as a question – repeatedly asking if a system being designed or used is 'safe, ethical and sustainable' won't immediately suggest all the details of the models I have alluded to, but it is a prompt and a route into those models.

Conclusions

I believe that safety engineering is at a crossroads. It needs to adapt to the complexities of current and emerging systems and to societal and environmental issues such as the impact of global warming. Some might argue that this is too big a change in role for the community, and I would view it as a target to strive for, not as an immediate objective. However, there is one immediate objective which I believe the community needs to adopt.

Bolt's "A Man for All Seasons" [11], focuses on the struggle between Henry VIII and Sir Thomas More, the chancellor, over issues of religion, power, and conscience. Whilst religion is outside our concerns here, being the 'conscience of power' and drawing to the attention of politicians not only the harms that complex systems can bring, but also their benefits, is something to which the community can, and I believe should, contribute.

I am pleased to see the Royal Academy of Engineering taking a lead in respect of the role engineering can play in reducing (the impact of) global warming. I see a similar model for safety engineers. If we consistently and persistently use the phrase 'safe, ethical and sustainable – treating it as a mantra (for all seasons) – this might begin to resonate with those in power and thus enable the safety engineering community to shape the future in a positive way, something we have perhaps not been good enough at in the past.

References

[1] G. Box, "Robustness in the Strategy of Scientific Model Building", 1978.

[2] J. Locke, "An Essay Concerning Human Understanding" (1689). Various reprints.

[3] L. Carroll, "Through the Looking Glass" (1871). Various reprints.

[4] M. D. Griffin, "How Do We Fix System Engineering?", 61st International Astronautical Congress, Prague, Czech Republic, 27 September – 1 October 2010, pp. 1-9.

[5] Aristotle, "Metaphysics", (See: W.D Ross et al. Aristotle's Metaphysics. Oxford University Press, 1925, pp 8-10.)

[6] "Safer Complex Systems: An Initial Framework", https://www.raeng.org.uk/publications/reports/safer-complex-systems (2020), accessed November 2021.

[7] National Transportation Safety Board. "Collision Between Vehicle Controlled by Developmental Automated Driving System and Pedestrian Tempe, Arizona March 18, 2018". (Published 2019).

[8] J.A. McDermid, Z Porter, Y Jia, "Consumerism, Contradictions, Counterfactuals: Shaping the Evolution of Safety Engineering", Safer Systems: The Next 30 Years, Proceedings of the 30th Safety-Critical Systems Symposium (SSS'22), 8th -10th February 2022, SCSC-170 (Published 2022).

[9] Maroochy sewage spill: https://www.risidata.com/Database/Detail/maroochy-shire-sewage-spill, accessed November 2021.

[10] J.A. McDermid, F. Asplund, R.Oates, J. Roberts, "Rapid Integration of CPS Security and Safety", IEEE Embedded Systems Letters (2018).

[11] R. Bolt, "A Man for All Seasons", Vintage (1990).

Image attribution
Lead image: ID 163537157 © Savagerus | Dreamstime.com. Lodz derailment: credit policja.pl

John McDermid Professor of Software Engineering and Director of the Assuring Autonomy International Programme (AAIP) at the University of York

John McDermid has worked on safety of computer-controlled systems for about 40 years and now leads the AAIP, focusing on the safety of robotics and autonomous systems, including those using machine learning. He has acted as an advisor to industry and government internationally and contributed to the development of standards. He is currently engaged on work on the safety and ethics of autonomous vehicles. He has supervised about 40 PhD students and published around 450 papers. He is a Fellow of the Royal Academy of Engineering and was awarded an OBE in 2010.

View From The Desk – 30 years of the SCSC

SCSC Inaugural Event

Tom Anderson and Joan Atkinson were key to the running of the Safety-Critical Systems Club for the best part of three decades. They often sat 'behind the desk', managing proceedings and ensuring the (mostly!) smooth operation of events. Tom and Joan reflect on the last 30 years of the Club and share some history of its formation and memories of the Club's activities and events.

A long, long time ago
I can still remember

These opening words to American Pie (Don McLean, 1971) refer to February 3, 1959, *the day the music died*, which was only just over 12 years earlier. In this article we will look back over the formation and activity of the Safety-Critical Systems Club, established in 1991 (with some reference to prior art going back to 1984) – so that's reaching back 30 years and more. As a result, to tell the truth, we don't now "still remember" lots of stuff.

Furthermore, if you're after good solid technical recollections of the evolution of principles and practice in engineering software-intensive systems for safety-critical applications, you won't find them here. Fortunately, the back catalogue of this Newsletter: *Safety Systems* and the proceedings of the *Safety-Critical Systems Symposium (SSS)*, published annually since 1993, comprehensively cover that deficiency. Indeed, in Volume 25, Number 3 of *Safety Systems,* you can read an excellent overview [1] of the first 25 years of the *Safety Club*, to use the familiar colloquial abbreviation.

SAFETY SYSTEMS
The Safety-Critical Systems Club Newsletter

25 Years of Safety Systems
and of the Safety-Critical Systems Club

Instead, we plan, basically, to gossip about those earlier times, as we watched (and, of course, shared in) the successful development of the Club; we hope you'll find some nuggets of interest in what is a somewhat discursive, and very informal, memoir.

How it all began

Way, way back, in the early 80s, concerns in industry and academia about the all too often highly unreliable behaviour of software, led to the formation of a national group (these days it might be labelled a focus group) of individuals which – after a pause for reflection – took the name Centre for Software Reliability (CSR). Needing a formal underpinning for this group, Bev Littlewood (at City University) and Tom Anderson (at Newcastle University) established two university research centres, also named CSR. The main focus at CSR (City) was on the assessment of software reliability, whereas CSR (Newcastle) concentrated on reliability achievement. This proved to be a timely initiative, since shortly afterwards the UK Government's Alvey Programme [2] drew active support on both of these topics from CSR.

An early CSR action had as its aims: to increase awareness of the need for more reliable software, to disseminate techniques for assessing and achieving it, and thereby stimulate improvements. The vehicle set up to deliver this was called *The Software Reliability and Metrics Club*, which created a Newsletter and a series of seminars (mostly one day, but some were longer); the inaugural meeting was held in London in October 1984, with over 100 delegates participating. The SRMC operated for just over two decades, but closed down after a total of 68 events – the final seminar was held in November 2005.

So now let's move on to the late 80s. Programmable electronic systems were by then moving rapidly, maybe too rapidly, into every sector, and the implications for public safety were becoming apparent to many. National awareness and concern led to a formal call – funded by the (then) Department of Trade and Industry (DTI) and the Engineering and Physical Sciences Research Council (EPSRC) – for proposed initiatives that could help to ameliorate the added risks that computers and, especially, their software could generate in embedded systems.

CSR took the view that, with national support, an upgraded version of the Software Reliability club could make a significant contribution in the safety arena. Responding to the call demanded a substantial proposal document; as usual this was completed with frenetic effort as the submission deadline approached. [We cannot forget Joan faxing long supplementary sections of text, prepared by Robin Bloomfield, out to Tom's hotel reception desk in the USA – yes, by fax, onto continuous-roll, heat-sensitive paper – cutting edge technology!]

After a competitive presentation, a contract was awarded for the formation of a *Safety-Critical Systems Club* (formally awarded to the BCS, on behalf of BCS plus IEE, with CSR to receive all the funding and do all the work). Financial support was tapered over three years, with the Club to meet specific targets and be self-sufficient when support ended.

The very first Safety Club meeting was held in July 1991, as a component of a DTI conference in Manchester. The issues to be addressed by the Club were seen to be a key concern for the conference delegates – 256 attended this launch meeting, requiring it to be switched out of the small room originally planned to the main auditorium. In fact, this remains the Club's highest attendance count (the next highest were 213 for the first SSS of the 20th century in 2001, and 205 for "Standards in SCS" – a 2-day event held in Cambridge in 1992).

At this point we would like to recall the dedication and commitment of the Safety Club's first "Coordinator": Felix Redmill. From the outset and for 25 years thereafter, the Club benefitted from his knowledge, experience, contacts and single-minded pursuit of the best possible event programmes, presentations and newsletters. Only ill-health now prevents Felix from continuing to intervene in the interest of the Club's objectives. However, we have been very fortunate in subsequently gaining from the new ideas and approaches delivered by his successors – Claire Jones, Chris Dale and the current enthusiast: Mike Parsons (from 2014 onwards). And in August 2016 we gratefully handed over responsibility for managing the Club to Tim Kelly, working with Alex King, at the University of York.

Some facts and figures

By January 1995 our membership database held contact details for around 2,500 members; all were recorded as individuals, but about 100 were included as part of a corporate package with their employers. [The corporate arrangement provided fully paid membership status at a group discount – the reduction in revenue for the Club was offset by the opportunity to renegotiate the group packages annually.] By promoting these group packages we increased the number covered to about 700 over the next few years, and although it then slowly reduced (to around 550), we were able to bring it back to over 700 by 2015. However, the non-corporate individual numbers declined to around 750 over this period (of course, many were now included in the corporate arrangements). A significant minority were non-UK colleagues, initially around 200, rising to 250 and then returning to 200.

Over the period 1991 to summer 2016 the Safety Club held a total of 86 seminars plus 17 tutorials, and the Symposium SSS '16 was the 24th in the series. The level of participation (speakers and delegates) was consistently encouraging, and supportive of Club finances. Our speakers rarely needed travel support and the average attendance count over the 127 events was 78.

The figures in the above paragraphs relate to Newcastle's period of responsibility for the Club; membership connections are now well in excess of 4,000; the current grand total of public events held (end of 2021) is 157, and that does not include the very many, rather more focused, Club working group meetings.

Some lessons learned

Of course, anyone involved in an activity that goes on for 25 years ought to gain something in understanding and experience, and – ideally – improve in capability. We are confident this applies in our case, given the number of occasions on which we needed to follow the swan's example: furiously paddling out of sight while trying to look serene on the surface. The following list tries to indicate some of the areas where we hope that we improved over time, though we quite often may have failed to look serene.

Event planning: appreciating the scope and scale of what needs doing, including comparing venue options (room capacity, facilities, location, availability, flexibility, cost), selecting, negotiating the rate, and finally booking (we learnt to just ignore the minutiae of hotel contracts, just like agreeing to software conditions of use, life's too short).

Event arrangements: paying close attention to details, including specifying room layout, catering requirements, tell the venue at least three times what the schedule is (with a hard copy on arrival – even then it will, occasionally, be ignored).

Event operation: crucially, of course, take the bookings and process the payments, but also organise badging, delegate listing and any other hand-out materials, monitor no-shows, last minute bookings, and unexpected arrivals. SSS is rather more demanding, with delegate bags to be filled with a considerable variety of enclosures (and not all bags are the same), plus providing assistance to our much-appreciated exhibitors.

Event venue costs: keeping these as low as possible by juggling the numbers. Very early on we decided that the best option was to pay on a per capita basis (hotel jargon is DDR – day delegate rate). But then the venue insists on a "minimum guaranteed number". So the aim is to achieve the smallest commitment for as large a room as might be needed, based on our own best attendance estimate. We used a sophisticated prediction method [Felix, Joan and Tom each made a (informed) guess, and we took the average]. Specific strategies were developed for SSS to cover (i) numbers at the banquet and (ii) bedroom accommodation – note that it would be sub-optimal to simply use the booking information supplied by our delegates; our aim was to have a place for everyone who actually turns up, but not to pay for any extras whatsoever – tricky! It's worth acknowledging that the Royal Marriott in Bristol gave us excellent support with this, but that elsewhere we sometimes struggled.

Event location: accepting national and logistic realities. In the early years of the Club it seemed appropriate to offer a wide spread of geographical locations, but we slowly recognised that London is indeed the centre of the UK. [An event we organized in Scotland attracted 50 delegates, but the vast majority were from England and they grumbled (a little).] Initially though, we avoided central London's inflated charges by selecting towns just outside the capital (e.g. St Albans or Woking). Delegates made it clear that this just made their travel more arduous, adding a suburban journey after they had reached London. And so, the Club's one-day events are now focused on London's city centre.

Club finances: identifying what really mattered. We realized that although pruning and optimising our costs was, of course, worth doing, the key concern in maintaining a break-even financial trajectory was income. Costs were predictable, but income was not. The previous section indicates how we sought to stabilize direct membership support by means of corporate package deals; we greatly appreciate the contribution of so many colleagues in industry who helped this to succeed. Income from one-day events barely covers their cost, so we focused on SSS. As a much larger event, running over three days, margins are more easily covered, and we developed the exhibition element as a very helpful income supplement. Our exhibitors, and especially the regular participants, deserve a vote of thanks for their ongoing support.

SSS evolution: the Club's flagship event. The annual symposium has always been a 3-day event, but the initial format was a tutorial day followed by two days of invited presentations (delegates could choose to attend either, or come to both). In 2012 the format changed to three days of presentations; in 2013 an even more significant change was made by selecting most of the presentations based on submitted abstracts. Adding an exhibition element was a further, highly beneficial, development – and not merely the financial support already mentioned. The exhibition reinforces the industrial focus of SSS and provides the ideal combination of mutual relevance: the services and products are directly relevant to most of the delegates, and most delegates are thus potential customers.

And lastly, we learned that after four intensive days (and evenings) at SSS, we were always somewhat drained (technical term: "knackered"), but found that a wee drink in the bar acts as a restorative – every time!

Some clear successes

Well, perhaps the most basic indicator of success of an organisation is survival. We thoroughly enjoyed looking after the Club until its Silver Jubilee in 2016, and are delighted to be anticipating the Pearl Anniversary of SSS in 2022.

Our personal perspective is necessarily subjective, but here is a summary nevertheless. The operational ethos has always been somewhat artisanal, associated with (but not a part of) the establishment, volunteer led and aided by largely volunteer effort – but always striving for a professional delivery of services and activities. We wanted to achieve truly face-to-face events offering genuine "networking opportunities"; a real meeting-up of like-minded safety personnel, thereby cultivating and building an interconnected "joined-up" community. And to be very welcoming, especially to new and younger colleagues (note the Club's current Safety Futures Initiative [3] to reinforce this) since clearly that is valuable to old hands and new faces alike. The characteristic manifestation of this was consistently demonstrated during the coffee and lunch breaks, which were invariably accompanied by a real (and therefore noisy) buzz of interaction. All in all, the fostering of a ***club*** of safety professionals that has now lasted for 30 years, keeping people in touch (pre-dating social media!).

The Club newsletter Safety Systems should certainly be mentioned here; indeed, the newsletter deserves far more than just a mention – so instead we refer you to the volume of selected articles "30 Years of Safer Systems [4]" (and the earlier edition "25 at 25" [5]) and also to the extensive repository of past articles available at the SCSC website (https://scsc.uk/Newsletter). The Club website has, in recent years, become a major repository for Club information, the primary vehicle for publicising events and activities, and an effective infrastructure for event bookings and membership registration. We gratefully acknowledge that this has only been possible thanks to the sustained efforts of the Club's webmaster, Brian Jepson, shown here seeking further inspiration, with dedication, through libation.

With some risk of hubris, we can surely include SSS in this section. The annual Club symposium is now a standard entry in early February in many calendars. It should be acknowledged that the fundamental contribution of the Symposium comes from the presentations and their recording in a published volume each year. A huge appreciation of the massive effort contributed in this way, by so very many individuals down the years, is entirely appropriate here.

The Symposium has also delivered an essential element of ongoing financial support for the Club's continued existence, via two mechanisms: directly, from the registration fees paid by delegates, and supportively, through the contributions made by our exhibitors. To enhance the experience of delegates and exhibitors, and to maximize footfall at the stands, we augmented conviviality by providing carefully selected fine beverages on each stand, adhering to a theme (that's right, initially malt whiskies, but subsequently beers and then ciders – and always the finest examples that we could identify using our networks of expert contacts); this innovation certainly seemed to go down well.

A further element of cordiality is offered each year at the Symposium "banquet", which always aims to offer good food, good wine, and good company. And also a little erudition: words of wisdom from an after-dinner speaker. We won't mention any names, but a soaring speech from an Air Marshal, and the verdict of a High Court judge (he's now a Justice of Appeal!) have featured. [The standard may have slipped a bit for 2022!]

Some problems encountered

You might naïvely think that with practice and experience and careful planning: what could possibly go wrong? Well, of course you wouldn't think that.

Although SSS gave us the most satisfaction, it also generated the most problems. And the one that occurred most often, and caused the biggest headaches: conference materials missing at the venue. We learnt that the best tactic for essential event materials (badging, programmes and delegate lists) was to carry them with us. [We learnt this the hard way, by having to create hand-made badges the evening before an event, handing them out to delegates with string for a lanyard. Not quite meeting our professional aspirations.]

Specific examples, of lost items, arose at the Belfry when the SSS proceedings were not delivered (we had to mail them out afterwards) and at the Brighton Metropole when all of our couriered boxes were handed over to the organisers of the preceding event, and carefully locked away in a "secret" cupboard. These were found only after following up with the courier company, then the courier driver, and then eventually contacting the organisers of the weekend event. Nightmare!

With disappointing regularity, and at various hotels, packages that we had very carefully labelled and shipped, and that had been delivered successfully, entailed lengthy searches by concierge staff before eventually being handed over.

As mentioned earlier, hotel bedrooms for a residential conference have to be guaranteed by us. The last thing we wanted was to have to pay for rooms that were not actually needed, and so – occasionally – we would be short by one or two bedrooms. A discrete request to a friendly and helpful delegate to stay nearby provided a simple solution. However, we recall two occasions when we were holding rather a lot of rooms less than needed. The first time this happened was when we were at the Belfry at the same time as Birmingham's Spring Fair (the hotel became fully booked and would not expand our allocation). We asked a dozen delegates to relocate to the very attractive New Hall hotel nearby, laid on transport, and covered the bar bills. Sorted. And just once, at SSS '01, the Royal Marriott could not help out, and we were six bedrooms short. We were very grateful to the RAF delegate contingent who agreed to stay as a group at the Bristol du Vin. (We knew better than to offer to pay for that bar bill!)

So, as you may have realised, our goal was to conceal any organizational problems from most, if not all, of our delegates. But here's one where that was just not possible. It was day 2 of SSS '02 at Grand Harbour, Southampton. Our presenter was just getting into his stride when the P/A system burst forth (very loudly) with music and announcements from a keep fit session elsewhere in the hotel, due to a misguided sharing of radio frequencies. Only a frantic search for a technician could fix that one.

"At the Belfry one organiser's back gave out … he left the hotel by being wheeled out to the car park sitting on a chair mounted on a hotel porter's luggage trolley".

Attending well over 100 Club events requires rather a lot of travelling – so some travel problems were inevitable. Here are a few anecdotes.

At the Belfry one organiser's back gave out (yes, it was Tom). He left the hotel by being wheeled out to the car park sitting on a chair mounted on a hotel porter's luggage trolley. Fortunately, there were very few spectators!

A Club seminar on formal methods in Peterborough (March 1987) had a splendid booking level of 126 delegates; the meeting room overflowed into the corridor! All in all, a good day. Since the venue was located directly across the street from the railway station, there was time after the event closed for swift refreshment in the bar before hurrying across to catch the train at 1800. But, oh dear – a major delay and the train was now due at 1850. Clearly the only acceptable option was to go for another pint and then back to the platform at 1845, where the rear lights of the departing train were still just visible, receding in the distance. The next one was due at 1930 so we stayed in the station. It eventually arrived just before 2000, and then was delayed again at York. We finally reached Newcastle at 2305, long before 'delay repay' was introduced.

In 2010, the Club (and CSR) operated the large Environmental and Safety Assurance Symposium event for MOD at Abbeywood, Bristol. That was the year that an eruption of the Eyjafjallajökull volcano in Iceland sent clouds of ash and dust into the atmosphere; the main impact on aviation was in April, but a second wave (as we now call them) in May meant our return flight was cancelled. We switched to a direct train to Newcastle from Bristol Parkway. It was rammed; Joan stood until Derby; no seat for Tom until Leeds.

However, our worst returning "rail" journey was caused by a closure of the East Coast Main Line between York and Darlington. Passengers waited in huge queues at York while coaches were, ever so slowly, brought in to transfer us all north to Bank Top station. Joan was frozen (stiff, then solid, she said); indeed she still complains about it now. Quote: *"I said we should have gone for a ****** taxi!"*

We were once trapped in London for an extra night. Very heavy unanticipated snowfall meant no trains or flights were operating at all. We only realised this rather late in the day and most hotels were, by then, full. Joan rang the massive Forum hotel (since renamed), and we managed to book two of the last five rooms.

A major snowfall in February 2009 had us worried about SSS that year, in Brighton. The conference team were at Newcastle airport ready to fly to Gatwick, with bags checked, when serious delays were announced. We were about to try to retrieve our luggage to see if we could head south by rail instead, when a late take-off was promised. At Gatwick, the only trains available were the ones we needed – trapped on the section south to Brighton. On arriving at Brighton station there were no taxis (because all of the local buses had stopped operating due to some snow on the roads). However, after waiting 45 minutes, a brave taxi driver picked us up. Given the problems we had had, and with bad weather continuing, we were seriously concerned about the risk of a low attendance. In fact, there were only about five no-shows. A special commendation is due to the tutorial presenter that year, Nancy Leveson. She had flown from the USA into Heathrow on the Monday, and just kept taking trains that gradually got her nearer to Brighton. By a very circuitous routing she eventually arrived at the hotel around 11pm. Indeed, we concluded that the only people who don't (eventually) get to SSS are those who don't set off.

So, let's end this on a positive note. We've massively enjoyed supporting the Club, and anyone who travels can recount the difficulties that sometimes arise. And although we may often have stayed in rather ordinary hotel accommodations, there have been splendid occasions too. One of these was when Joan was allocated the Presidential suite at the Belfry (probably the best room she's ever stayed in). And to add to the joy, we overheard a very wealthy gentleman from overseas complaining at reception because he could not just walk in and get a room: *"I'll pay for the Presidential suite" "I'm afraid it's occupied, sir"*.

But best of all, when we held SSS at The Grand at Brighton, your authors were allocated (at no extra cost!) almost the entire first floor frontage of the hotel (the Thatcher suite, we called it). The layout was: huge bedroom, huge lounge, small dining room, huge lounge, huge bedroom. Although the dining room was not included, we hired it personally for the night before the conference, opened up all five rooms (just to show them off) and hosted a private dinner for eight. A most memorable evening.

Ah well, go on then, just one more problem scenario. The organisers arrived at the SSS venue hotel on the Monday, at around 11.00, only to be told by reception that no bedrooms had been reserved for us, nor for any of our residential delegates. Just picture Joan's reaction. Speculate about what she said. Rather a memorable morning, actually.

References

[1] Safety Systems, Volume 25, Number 3, May 2016, Felix Redmill, https://scsc.uk/scsc-144

[2] The Alvey Programme, https://en.wikipedia.org/wiki/Alvey, accessed January 2022

[3] The Safety Futures Initiative, Zoe Garstang, https://scsc.uk/gf

[4] 30 Years of Safer Systems: Three decades of work in the field of safety-critical systems as told through the SCSC Newsletter, Louise Harney, Mike Parsons, Paul Hampton, Roger Rivett, Wendy Owen (Eds.) https://www.amazon.co.uk/Years-Safer-Systems-safety-critical-Newsletter/dp/B09KNCYKDL/, October 2021.

[5] 25 at 25: A selection of articles from twenty-five years of the SCSC Newsletter Safety Systems, Mike Parsons, Graham Joliffe, Tim Kelly (Eds), https://www.amazon.co.uk/25-selection-articles-twenty-five-Newsletter/dp/154089648X/, January 2017.

Tom Anderson and Joan Atkinson

From 1991 to 2016 Tom Anderson directed the SCSC within the auspices of Newcastle University, where he was Professor of Computing Science. His research interests addressed fault tolerance and, more broadly, dependable systems (encompassing safety and security). In 1984 he established the Newcastle branch of the Centre for Software Reliability, which provided a supportive environment to a series of research projects, and also organised over 250 external conferences and seminars – all with a strong industrial orientation. From 1992-97 he was Head of Computing Science; 1998-2002 Dean of Science; 2008-2012 SAgE Dean of Business Development. Tom retired in 2016, but continues to be active in the SCSC Steering Group and maintains engagement in outreach via CSR Events. Thanks to Covid restrictions he has designed and scratch built a rather splendid garden shed.

Joan Atkinson joined Tom at CSR, Newcastle University in 1985 where she became the research centre's Administrative Coordinator which, as well as support for the centre's academics and their research, involved full responsibility for the administration of the SCSC. The events referred to in the previous paragraph were, of course, all organised by Joan – in fact there were 256 events altogether, total duration 422 days, with an average daily attendance of 86 (equivalent to looking after 100 people for a year). She too retired in 2016, and now does the work of CSR Events as a self-employed PCO (professional conference organiser). Despite Covid restrictions, as Chair of the Washington Village in Bloom group, she led them to victory in the Northumbria in Bloom competition (best overall entry) and was awarded a trophy cup only slightly shorter than herself.

THE SAFETY-CRITICAL SYSTEMS CLUB

Seminar: Managing 'Black Swans':

Handling Rare and Severe Events Now and in the Future

8th April 2022, London, TBC hotel and blended online

Bookings at:
www.scsc.uk/events

This seminar is an opportunity to hear about management of rare and high impact events across different industry sectors and how this is likely to change in the future.

It will be useful for safety practitioners, safety managers, and for those involved in the planning and management of high-impact events.

Details at: www.scsc.uk

How to Manage Unexpected and Severe Events

www.scsc.uk

This seminar will consider how to plan for and manage recovery from 'Black Swan' events in a safety context These are events which are rare, unexpected and have high impact. Examples might be the Fukushima nuclear disaster or the loss of Malaysia Airlines flight 370.

There are many aspects to the management of such events including planning, preparedness and dry-runs of contingency processes. When an event occurs, it is necessary to quickly establish the nature and scale of the problem, stabilise the situation, prevent of a cascade of failures, assess risks, provide a contingency service if possible, communicate to all stakeholders and eventually recover normal operations.

Communication, obtaining reliable status information and rapid assessment of risks are critical but may be difficult. Hard data may be limited, and situational awareness, human factors, organisational experience and safety culture all come into play.

The first part of this seminar looks at the current position in various industries. The second part examines the situation when upcoming automatic and autonomous functionality is involved. How do we make risk-based judgements when human involvement is small?

There will be workshop session where delegates can explore the events and the possible solutions further.

The Future of Human Factors?

Human factors have always been important when considering how accidents may be prevented or evaluated. In this article, John Ridgway explores how such considerations may play out in the future. In particular, one has to consider how the factors that influence safety-related decision-making will be judged after one takes into account the methods and processes that are likely to be in place.

A palpable tension was hovering in the air like the lingering stench from a cheap e-cigarette. A court that had previously gasped from want of belief rather than fresh air, now held its breath as the Counsel for the Prosecution rose to her feet to deliver her closing speech.

"Members of the jury," she opened, "you have been witness over these last two days to a tragic story of woeful dereliction, matched only by an even more woeful failure to apply the ethical judgment that all members of the public have a right to expect from those given the responsibility of ensuring their safety. This was not a failure of a component, as the Defence would have you believe, but a failure that struck at the very heart of the decision-making process. A failure of decision-making that led to the tragic deaths of Mr and Mrs Cooper as they embarked upon what should have been a perfectly safe journey upon the Gender-Neutral Royalty's highway.

You have heard a truly bizarre attempt from my learned Counsel for the Defence to exonerate the defendant on the grounds of logicality. No doubt these arguments will be repeated shortly, but, as you listen to them, I ask you once again to consider how such a failure of judgment was possible. How can an operator instructed to set road signs that are vital for the safety of the road user, possibly justify setting them in such a way as to knowingly increase the risk to not one, but two, members of the general public? Normally when considering operator fallibility, one is confronted with unfortunate errors that are quickly regretted. But have you heard one note of contrition from the defendant in this court? No! Just an insistence that the twisted logic applied should be accepted as the optimum safety decision to be made under the circumstances.

Well, I'll leave you, the members of the jury, to decide upon that matter yourselves. However, I put it to you, that any right-thinking mind would look upon the decisions made on that fateful day and come to the inescapable conclusion that the defendant is guilty as charged; guilty not only of gross neglect but also of a reasoning that you must surely agree was most egregiously flawed. I thank you all."

As the murmurs spread through the auditorium, it was clear that the speech had gone down well. The court had been invited to understand that this was not a tragedy borne of physical frailty. Instead, the frailty lay in an inability, under stress, to apply a reason and rationality that would be recognisably ethical to those who were in a position to judge. Today, such a judgment was being made by twelve of the defendant's peers, and it mattered now, more than ever, that the Counsel for the Defence could make a strong enough case for believing that not only rationality and reason, but also ethicality and morality, had lain at the heart of the defendant's decision-making. A hush descended upon the auditorium as she rose to her feet and turned to the jury.

"The frailty lay in an inability, under stress, to apply a reason and rationality that would be recognisably ethical".

"Members of the jury, my learned Counsel for the Prosecution has succeeded admirably in fomenting righteous indignation, but I put it to you that emotions should be set aside when determining your verdict. I have carefully explained over these last two days the complex interplay between traffic congestion and road safety. I have ably shown that all usage of the Gender-Neutral Royalty's highway entails risk, and that this is a risk that each and every one of us accepts when we step into our vehicles.

Indeed, there are only two traffic states that are truly safe: firstly, when the congestion is at its minimum because the road is empty, and secondly when the congestion is at its maximum, resulting in a gridlock that has ground the traffic to a halt. I have also shown you the risk profile curves that demonstrate that the accident risk is at its maximum at the mid-point just as the laminar flow of traffic starts to break down and shock waves start to develop within the traffic flow. It is at this point of flow breakdown, the onset of chaos if you will, where one sees the occurrence of unexpected queues that are the main cause of rear quarter collisions. Furthermore, it is at such a point that the temptation to make dangerous lane changes is at its greatest.

Therefore, the operator in charge of setting signs has a duty to implement traffic management strategies that move the traffic away from this phase transition whenever possible.

Sometimes this will entail diversions to alleviate traffic flow on a particular stretch, but it may also require the increasing of traffic volume through judicious means, in order to quickly encourage a slowing of traffic. This is all that the defendant was doing on that fateful day. By setting lane change instructions that encouraged Mr and Mrs Cooper into a collision with another vehicle, two major benefits had been identified. The traffic behind the resulting accident would grind to a halt, and the road ahead would empty. This win-win situation was marred only by the sad deaths of the couple concerned. However, since the defendant had anticipated and avoided an even worse accident, who amongst us can say that the wrong decision had been taken? I put it to you that the decision was taken with safety optimisation in mind and that the decision was taken under the most difficult of circumstances. On such a basis alone, I must conclude that the only logical decision available to you is to acquit. I therefore appeal to your own respect for logic and ask that you draw this conclusion: By all reasonable judgment, the defendant is not guilty."

"The seeds of doubt had been successfully sown as the jury members were now being asked to place themselves in the position of the defendant".

With that, the Counsel for the Defence retook her seat before turning to her assistant. A small but discernible smile played across her face as she listened to the appreciative mutterings now echoing within the courtroom's small confines. The seeds of doubt had been successfully sown as the jury members were now being asked to place themselves in the position of the defendant. The crux of the matter was this: Had they found themselves in the same position, how would they have reacted? Put another way, how normal was the thinking of the defendant? Was this, at the end of the day, the only criterion that we should be applying to determine ethicality? As the auditorium continued to grapple with these questions, the murmuring grew ever louder, to the point that the judge felt it necessary to intervene.

"Silence in the court!" she bellowed. "Members of the jury, you have now heard the closing statements from both counsels. I ask now that you adjourn to make your decision."

By now the tension in the courtroom was barely tolerable, and so it was to everyone's advantage that the verdict was returned with the minimum of delay. As the court reconvened, the judge once more struggled to regain control.

"Silence! May I remind you that this is a court of law." Having thus re-established her authority, she slowly turned to the jury. "Members of the jury, have you reached a decision upon which you are all agreed?"

"We have."

"And what is that decision?"

"Guilty, M'lady."

Not for the first time, the judge had a crowd management issue to deal with as the auditorium erupted into loud cheering.

"I will have silence in my court!" she maintained with a well-judged *fortissimo*. Turning to the defendant, it was now the judge's chance to draw conclusions.

"You have been found guilty of the most appalling error of judgment, and for that there can be only one verdict. But before I pass sentence, I feel it only fair to draw attention to an important principle. So often in these situations it is the operator at the coal face that is brought before this court, and yet no accident can be said to have been caused by a single factor. In so many cases one has to take into account systemic failings that made the operator error all the more possible. This case is no exception and so it would be remiss of me not to point out that the operator would have not made this calamitous decision if the software engineer who had programmed it had ensured the inclusion of the necessary subroutines for checking compliance with all relevant ethical constraints. Consequently, I will be advising that a review be held into all future artificial intelligence programming, as employed on automated safety systems development for the Gender-Neutral Royalty's highway. In the meantime, I sentence the defendant to be immediately decommissioned and add that it shall not be re-commissioned until such a time as the safety case has been approved for the regression testing of its software upgrade. I would also like to thank the jury for its most sage judgment and I advise that you should all be excused further jury service until after your power units have been refurbished. This court is now closed."

Now, and only now, the excesses and exuberance of the auditorium were to be encouraged. A difficult judgment had been made in a manner that appeared to be to everyone's satisfaction. But as the clamour in the public gallery slowly subsided, it was only the most observant and alert amongst them who will have witnessed the Justices' Clerk leaning forward before deftly switching the judge into standby mode.

John Ridgway, Retired

Following 30 years in various quality and safety assurance roles, whilst working for a contractor developing traffic management solutions for both domestic and foreign clients, John is now enjoying a relatively uneventful retirement on the edge of the North York Moors. John would like it to be known that he learnt everything he knows regarding the UK's judicial system from watching poor courtroom dramas.

Planes and Computers

Stan Price has recently published his autobiography "Trains, Planes and Computers" chronicling his long career in systems safety, from designing, auditing and testing safety-critical systems through to research into making them safer and even acting as an expert witness in court. Stan shares some of his thoughts and insights from working in the discipline for over 30 years.

The Prehistory of System Safety

Even before systems safety, and in particular, the probity of software, became a specific topic, those involved in such systems, including myself, were well aware that systems could kill and maim. It was around 1968 when I was first responsible for a safety-related system. It was for the production of the Operating Data Manuals (ODMs) for the then current Manchester (AVRO) designed aircraft – the Nimrod and 748. The ODMs, as the name suggests, indicated to pilots how the aircraft could safely be operated and consisted largely of tables. For example, they indicated minimum runway lengths at particular take-off weights, airfield altitudes and ambient temperature. For each phase of flight, e.g. take-off, cruise, or climb, there were three programs in the process of producing the relevant part of the particular aircraft's ODM.

The first of these calculated engine performance, which was then used by the second program to calculate raw aircraft performance data. The final program in the suite sorted this raw data into the table that went into the ODM. This presented unique formatting problems, in particular, as the environment got more arduous – hotter and higher; the aircraft could not operate there, so there was no entry in that part of the table.

Obviously if the data in the ODMs was inaccurate and was acted upon, there was a danger that safety could be compromised. Indeed my ODM system came under suspicion when two aircraft slid off the same runway in the same afternoon. Fortunately for me, the problem was an inaccurate hand-produced correction figure for the landing distance required on wet grass.

A Move to Air Traffic Control

Late in 1972, I moved into the realm of Air Traffic Control (ATC), with its obvious safety implications, and was responsible for judging that a projected system's software was of such poor quality that safety would be again compromised.

It was a system for assisting controllers in knowing more exactly the sequence of aircraft landing at Heathrow. This involved calculating the timing and speed of aircraft leaving the four stacks so that the spacings when they landed on the runway were the minimum commensurate with safety, and hence its use was optimised.

A prototype of the system was being developed by the Royal Radar Research Establishment at Malvern. One of my software engineer colleagues, Dave Neumann, and I visited Malvern and viewed the system. We discovered the software was very poorly written. It had no structure and comments were also non-existent. Its documentation was also very limited. Dave and I therefore reported this, and the project was cancelled. Its quality was so bad that any deep consideration of safety was unnecessary. I feel generally that the quality of software has improved over the years since, and particularly, where relevant, with the focus on safety.

My major involvement in ATC systems was in the UK's acquisition of the US en-route ATC computer system (the 9020 Project). Even though the original system would not, in its US operational life, handle aircraft flying east of the Greenwich Meridian, the specification allowed for the possibility. But it appears that this was never tested before going operational in the US; but upon testing in the UK, it folded the country over at the meridian. For example aircraft flying over Ipswich were being shown as being over Bedford. Hardly safe, defeating the whole purpose of ATC – to stop aircraft colliding. Suffice to say this was corrected before the system went operational. The principal lesson to be drawn was the dangers of assuming that a system that was safe in one domain does not mean it will be safe in another.

An Expanding Role

Later, I was involved in a police command and control system. This would have the purpose of real-time allocation of police assets to incidents requiring their attention on a geographic basis. Misallocation could mean that some safety-critical incidents would not be attended to thus diminishing safety. My role would be a key one in choosing the contractor to implement the system and then oversee its installation.

Next, I was asked to monitor three projects in the DTI/Research Council Safety-Critical Systems Research Programme. I also believe the Programme spawned the SCSC Club. The projects were:

MORSE – A Method for Object Re-Use in Safety-Critical Environments with partners University of Cambridge, Lloyds Register, West Middlesex Hospital, Transmitton, Dowty Controls and British Aerospace Airbus

SPAM – Investigating Security Paradigms Validity for Safety-Critical Environments with partners EDS-Scicon and Lloyds Register

PRICES – Productivity, Integrity & Capability Enhancement for Software, and Human Factors in Safety-Critical Systems Development involving Open University, City University, Lloyds Register, Rolls- Royce, Bae SEMA, G P Elliot Electronic Systems and Analysis Consultants.

By this time (1994) the pro-active consideration of safety using techniques such as HAZOPS etc had become much more structured and mandated replacing the simple "does it meet the spec" criteria of yesteryear.

What goes on out there?

Subsequently I performed a coordination role within the Programme including two initiatives, which I conceived and got funding for. One of these initiatives was a series of workshops devoted to specific topics, which I considered, were of key relevance to successful and particularly safe systems. My judgement on what was key was heavily influenced by my previous industrial experience. To make the workshops manageable, the number attending was restricted to twenty or less, split equally between academia and industry, largely but not exclusively, drawn from participants in the safety-critical programme, participants that I rated.

I chaired the workshops, and each had a rapporteur who produced a report chronicling the proceedings and the conclusions. This was circulated to the attendees for their comments, and a final version incorporating these was then published. The workshops were opened by myself with an introduction that cited the purpose of the workshop and its format.

The latter, after the introduction, consisted of presentations from the safety-critical programme projects that were relevant to the topic of the workshop. These were followed by comments on the presentations from the so-called catalysts (chosen to stimulate discussion), which led into general discussions followed by a summary.

The initial workshop, under the title, 'What goes on out there?' addressed the gulf between what the research community thought happened in day-to-day industrial/commercial practice and what actually happened. As well as the DTI sponsorship of the workshop, it also ran under the auspices of the SCSC, and over the years, I also made contributions to several of its events.

"... the deliberations at these workshops and their proceeds had a degree of influence particularly in the hitherto neglected topics of Human Factors and Data."

The other workshops that followed the initial one addressed the following, in relation to the safety of systems:

- Human Factors
- Software Requirements Elicitation and Capture
- System Assessment
- Artificial Intelligence
- Process Models
- Data

I believe that the deliberations at these workshops and their proceeds had a degree of influence particularly in the hitherto neglected topics of Human Factors and Data.

Expert Witness

My final professional contribution to the safety-critical world came as an expert witness in a court case around 2001.

Electronic Data Systems (EDS) was contracted to supply to the Civil Aviation Authority (CAA) a new computer system at Prestwick in Scotland to support the control of air traffic over the

North Atlantic. The CAA cancelled the contract after the design milestone, citing non-performance on the part of EDS and EDS took the CAA to court for some forty-two million pounds. Because of my air traffic control and computer background, I was approached directly by EDS's solicitors to be an expert witness for them.

Curiously, this was not so much for my system development expertise; a team of three other experts was handling that, but on the safety issues in the case. EDS's legal team believed that the CAA might play the safety card. Unfortunately, my other commitments at the time meant I could not really take up the assignment, and at first, I said no. However, the fee I was eventually offered, a four-figure day rate, plus an agreement that I could employ a researcher, made me change my mind.

The researcher I employed was John Smith, who had been the Plessey manager involved in the 9020 Project during my days with the Civil Aviation Authority, and who had subsequently produced the safety case for the London Air Traffic Control Centre.

Insofar as John's and my inputs to the case were concerned, they were significant in two respects. Firstly we smoked out the fact that the first part of four of the CAA's safety procedures amazingly did not exist, and therefore, their criticism of EDS's safety processes in the case was therefore unrealistic, to say the least. Secondly, in my initial report, I had pointed out that there was no common agreement on how safety should be built into software systems, so their criticism of what EDS was doing was not necessarily valid. The CAA retorted that there was agreement and cited numerous standards. To this, I asked the simple question: if there was a common approach, why was there a need for so many standards, which in some areas were even contradictory.

"... if there was a common approach, why was there a need for so many standards..."

Are We There Yet?

This may not be the situation now, but as a distant observer in retirement of the safety-critical systems scene, I am amazed that unsafe systems are still going operational. Two significant ones that come to mind are the Boeing 737 MAX [2] and the so-called smart motorways [3]. It may well be, at the technical level, that we are now much better at ensuring safety, but until we stand up to the financial and political pressures sometimes put upon us, our detailed work will come to nought.

References

[1] Trains, Planes and Computers, https://www.amazon.co.uk/Trains-Planes-Computers-Executive-Pass/dp/1802271252, Stan Price, 2021, ISBN: 978-1-80227-125-6 and 126-3

[2] Boeing 737 MAX – Safe to Fly? Paul Hampton & Dewi Daniels, Safety Systems Volume 29, Number 1, Feb 2021 https://scsc.uk/scsc-162

[3] How Smart Are Our Motorways? John Ridgway, Safety Systems Volume 28, Number 2, May 2020, https://scsc.uk/scsc-158

Top image: AVRO748 undertaking rough airfield trials at RAF Martlesham Heath now BT Research.

Stan Price spent over 40 years developing, evaluating and researching IT systems – many of them safety-critical principally in the Aviation sector. He is a Chartered Engineer and Member of both the Royal Aeronautical Society and British Computer Society. He has performed visiting posts with Sheffield and Salford Universities and given evidence in over 40 court cases involving IT.

The SCSC and the Internet (actually the World Wide Web but...)

The SCSC website has become an essential resource for members, not only in providing information about the club and upcoming events, but also in providing a wide range of other services such as access to publications, working group resources, multi-media material and has been critical in ensuring the club could weather the Covid-19 pandemic by facilitating streaming-based services. Brian Jepson, the SCSC webmaster, describes the history of the website from its humble beginnings and how it has evolved over the years.

The Safety-Critical Systems Club has a history that runs parallel to that of the World Wide Web (WWW). These days most people use Internet, which has been around since the early 1980s, interchangeably with the World Wide Web (WWW), designed by Tim Berners-Lee, that operates as a service using the internet. Most of what I'm talking about here is WWW but 'Internet' makes a snappier title. Both the WWW and the SCSC appeared to the world at the beginning of 1991, but it took a few more years before the Club started to use the web in earnest.

In today's world, the Club could not survive without a website. In the year from mid-2020 to mid-2021, the website, in conjunction with email, Zoom video conferencing, digital publishing on Amazon, and video streaming through YouTube, provided online seminars, a three-day symposium and allowed the working groups to continue their work.

The early days

From its inception in 1991 through to 1997 the Club had no presence on the web. The Centre for Software Reliability (CSR) at Newcastle University was running the Club with operations being conducted by telephone, post and email.

www.csr.newcastle.ac.uk

The screenshot here shows an early CSR web page from 1998.

At this time the website only provided basic information about the club, its objectives and what it does.

These pages were hand coded in HTML and were difficult to maintain. In later versions the technology adapted to include a database, CSS and server- and client-side scripting but remains focused on substance and accessibility over style.

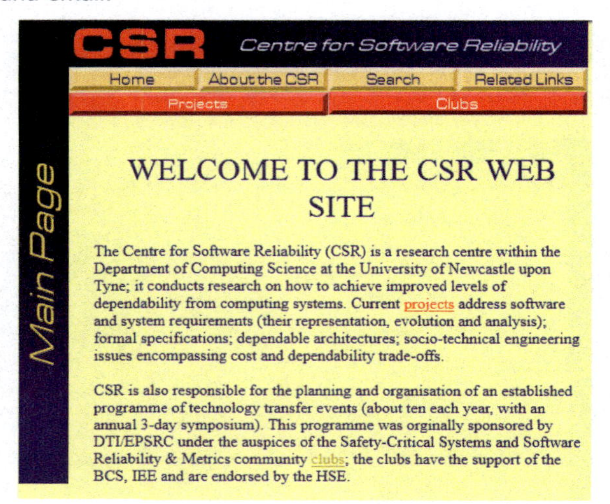

www.safety-club.org.uk

In 1999, the www.safety-club.org.uk domain was registered to give the Club its own identity, though this was the existing information served by the Newcastle University using the CSR Clubs pages.

This screenshot shows an updated CSR page from 2001 accessed as www.safety-club.org.uk that focuses on the Club and includes the next Club and related events with linked pages containing details of each event. This style of web page remained in use through to February 2008.

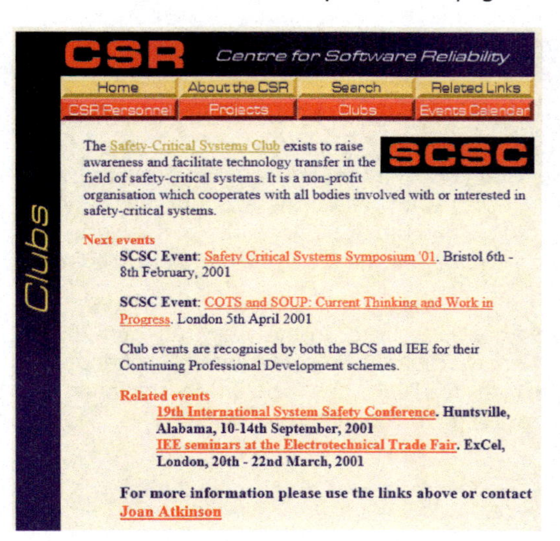

www.scsc.org.uk

By 2004 the Club was increasingly being known by its SCSC initials, and luckily, the www.scsc.org.uk domain was still available, so this was registered as an alternative to the full safety-club domain.

It would be difficult today to get hold of a short domain name like this and when, in 2014 ,the abbreviated .UK domains were introduced we exercised out right to also acquire scsc.uk giving us a seven-character identity.

It was also becoming difficult to maintain the pages hosted by the University, so www.scsc.org.uk was kept separate and developed as a new website. This new site remained, linked to the CSR website, but it was able to rapidly expand to include more useful information such as back issues of this, the *Safety Systems,* Newsletter as visible below. This version of the website made use a database to store all the resources and scripts to dynamically generate pages as requested allowing much easier and quicker updating.

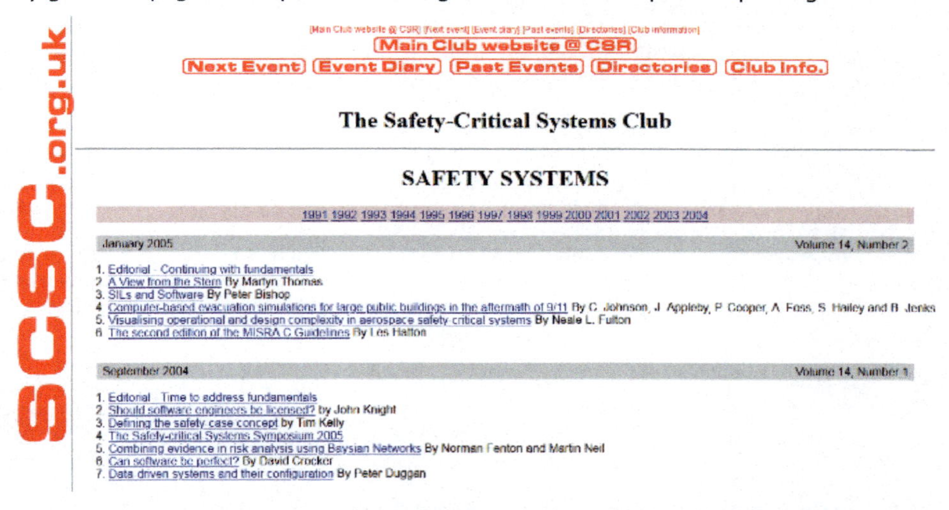

In February 2007 the CSR pages hosted at Newcastle were abandoned and both domains now delivered the new website which had developed to include more publications, a directory of tools useful in the development of safety-related systems and information about club membership. The screenshot below shows an early version of this new website. Note the logo as a flag on the red flagpole that has now become an SCSC style used on all pages and many handouts.

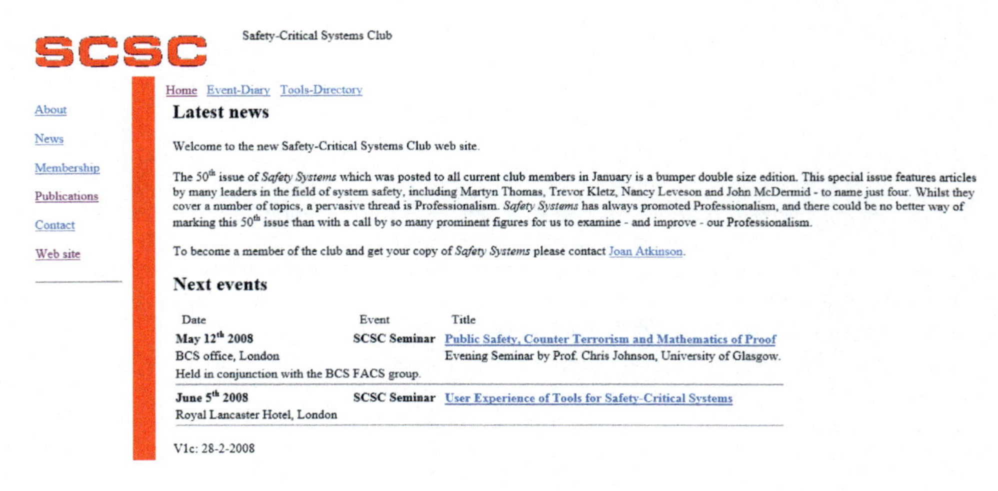

Recent years

In 2014, the Club acquired the shorter www.scsc.uk domain and, to reflect the international appeal of the Club, also acquired www.thescsc.org. All the Club domains lead to the same content from the canonical www.scsc.uk website.

www.scsc.uk
www.thescsc.org

The website has now grown to include a comprehensive history of club events with over 1400 resources available to Club members including presentation materials from events, books, working group guidance documents, symposium papers and 69 issues of this, the club newsletter.

There are ten *working groups* supported by the club covering topics ranging from Autonomous Systems Safety through to Safety Culture and The Safety Futures Initiative. Each working group has its own web pages where ongoing work can be shared.

There is a *publications* area where members can download digital version of the symposium proceedings, newsletters and other documents such as the guidance produced by working groups. There is also a *community space* forum style area where thoughts and opinions can be shared.

In the *Catch up* area are details of many of the Club's past events. Most presentations at events are now recorded, and the videos of these are available for members, together with copies of the presentation slides and any other handouts. Because of the Covid-19 pandemic, events between mid-2020 and mid-2021 have been online only, and, since October 2021, as hybrid events with both in-person and online participants.

Brian Jepson, SCSC Website editor, 2004 onward.

Brian has 38 years' experience in software and system safety in the defence sector but has now retired and spends his time supporting the SCSC and restoring a Land Rover 101.

The Future of Safety Engineering and Assurance

The SCSC has come a long way since its inception 30 years ago, and has achieved a great deal throughout that time, undoubtedly contributing to making systems safer. Over those three decades, the club has needed to adapt to a changing world: the use of technology in society has expanded rapidly, and systems have become more powerful, complex and distributed with many new disruptive technologies (such as AI) making safety assurance ever more difficult. So what will the world be like in another 30 years? What will the concerns of the club's members be in 2052, and what achievements will the club be celebrating?

Jeremy Messersmith, in his ukulele song "Everybody Gets A Kitten" [1], offers his particular optimistic vision of the future:

"Gotta say the future's awesome, everything is a-okay!
All the work is done by robots, every day is Saturday.
Future people all have jet-packs, fly around in flying cars..."

And, as hinted by the title, he goes on to predict that:

"Everybody gets a kitten, a new one every single day ...
You can name if you want, or you can give it away!"

Jet packs and flying cars? Probably – we have prototypes of those now; but, as we've seen, the logistics of distributing vaccines to 50+ million citizens in the UK alone has been immense; imagine the distribution infrastructure, processes and personnel required to ensure everyone had a kitten delivered *every day*... well, perhaps not.

To get a, hopefully, better informed and sagacious answer to these questions, members of the SCSC Steering Group were canvassed for their opinions on how they see the future of safety engineering. The following summarises some of their predictions in response to four specific questions.

The pressing concerns of the day might be best illustrated through the title of the key note speakers' talks at SSS'52, but what might these be?

The continued progression of autonomous systems featured in a number of suggested titles:

- Why my iPartner isn't always pleased to see me
- Training Adaptive Systems of Systems in a Secure and Ethical Way
- Why Manually-Driven Vehicles are a Danger and Should be Banned
- The Importance of Non-human Factors in Safety Assurance
- Can AIs Argue Their Own Assurance?

There are also views that system safety will be an increasing concern for systems operating off-world in space and on other planets:

- Fatal Mars Rover Collision in 2050: Final Accident Report and Analysis
- A review of safety standards for commercial space transport vehicles

Other titles suggest that safety assurance challenges will emerge from novel technologies:

- How Weather Control Failed in the Storms of '51
- Regulating System Safety in the Metaverse [2]

The expansion of safety consideration from primarily focussing on harm to individuals and the associated technical mitigations to include the wider societal and environment impacts, and encompassing the trade-off between harms and benefits, is expected to bring more expansive concerns:

- Safety, Ethics and Sustainability of Domestic Space Flights
- Safety Assurance of Earth's Digital Twin

As with our existing standards – some being in use for several decades – standards are expected to continue to feature in the future:

- The key differences between IEC 61508 Editions 9 and 10

Autonomous Systems and Artificial Intelligence (AI) are some of the current bêtes noires for Safety Engineers, but what sort of technologies will be the most challenging for safety engineers in 2052?

The challenges of AI are still expected to be present well into the future, with new developments confounding the assurance progress that might have been made in the meantime:

- Explainable systems that make up false explanations for their decisions, i.e. create lies
- The automation of safety certification using AI based on 2040's practices and how to get new techniques accepted
- Evolutionary systems that breed new behaviours
- Personal robotic assistants

As with the key note speaker title, space travel is also seen as a future challenge as the capability becomes more accessible and available to a domestic market. However, this is just one example where technological expansion will need the safety community to take a wider view, such as: what are the impacts on global resources, climate change and risk distribution from such activities?

Challenges in the Healthcare domain and medical science also features as presenting future challenges, both technically and morally. Some Covid anti-vaxxers already believe the technology exists to inject microchips into people. When such technology is available, how will we deal with the safety implications and need to weigh the risks of not introducing (life-saving) technology? Specific examples are:

- Healthcare Nanobots
- The embedded man-machine interface – cyborgs
- Robotic surgery

Other suggested areas that will present challenges are related to the challenges we face now:

- Highly adaptable, configurable systems – just what are they doing today?
- Systems of Systems; Systems of Services – everything interconnected and inter-dependent

And of course, we will still have the same issues that we've not been able to solve in the last three decades, such as the metrics used to measure software.

Interestingly, there were not many responses in terms of the integration of safety and security disciplines; only some concerns around Ubiquitous Communications and Computing, but perhaps this reflects an optimism that safety/security integration will be eventually 'solved'...

What new tools, techniques and methodologies will be available to safety engineers in 2052?

New tools and techniques are certainly anticipated, but there is also an expectation that the tried and tested techniques will also still be with us. Firstly, new suggestions:

- Better visualisations
- Simulation/animation models
- Virtual environments
- Quantum Risk Assessment
- AI-prediction in Safety analysis - now we don't have to guess 'What if?

And evolutions of the more familiar standards and techniques that we have now:

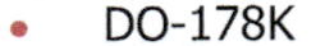

- DO-178K
- STOMP: The Systems-Theoretic Outcomes Model and Processes (c.f. STAMP [3])
- Safety V (c.f. Safety I and II [4])
- Dependable software reliability techniques

There is also an acknowledgement that existing tools have limitations (eg. They tend to focus upon single failures) and will find it difficult to cope with more complex systems. It's therefore expected that new tools/techniques will be required to enable safety analysis of increasingly complex systems without incurring disproportionate time or cost.

And finally, what club achievements will we be celebrating over the previous 30 years?

On a fundamental level, one basic achievement will be for the club to still be going strong in another 30 years' time and still fulfilling its community objectives. The Covid pandemic has been a challenge to the club financially, and has demonstrated how much it depends on its members and in the knowledge-sharing environments where it thrives.

As well as having much increased membership from all around the world, a demographic of much younger members is anticipated, with the range of safety concerns also expanding in a more intersectional way, to include areas such as environment, sustainability, ethics and inclusion.

By expanding the scope of safety to include ethics, sustainability and the wider societal benefits and impacts, the club will have played a more influential role in government policy making and contributed to tackling the world's bigger issues, such as global warming and inequitable risk distribution.

Embracing new media formats is also expected; the enforced move to online events in recent times has shown that multi-media events can work, and this will only get more interactive with the Metaverse, Virtual Environments and Augmented Reality.

And to conclude, might we be having our first ever successful meeting held in space or even on a different planet?

References

[1] "Everybody Gets A Kitten", Jeremy Messersmith, from the album "11 Obscenely Optimistic Songs For Ukulele: A Micro Folk Record For the 21st Century and Beyond", 2017

[2] Metaverse, https://en.wikipedia.org/wiki/Metaverse, accessed January 2022.

[3] An Introduction to STAMP, https://functionalsafetyengineer.com/introduction-to-stamp/, accessed January 2022.

[4] The Safety-II approach: Learning from what goes well, https://www.patientsafety.com/en/blog/safety-2-versus-safety-1, Jens Hooiveld, accessed January 2022.

Article by Paul Hampton SCSC Newsletter Editor with thanks to our Steering Group contributors: Mike Parsons, Brian Jepson, John Spriggs, Graham Joliffe and Tim Kelly.

Can We Quantify Risk? Event Report

The "Can We Quantify Risk?" seminar was held on 21st October 2021 at the Radisson Blu Edwardian Bloomsbury Street Hotel, London and virtually online. This was the first in-person event held by the SCSC in over 18 months and the first to be a blended event with delegates also attending online. Mike Parsons, chair of the seminar, reports on the event and assesses how well the blended format worked.

The day opened with Mike explaining how pleased he was to see delegates in person at an SCSC event! He said that the Covid-19 pandemic had made us all risk estimators to an extent. He then introduced the event mentioning some aspects of risk and new areas for discussion such as autonomous road vehicles.

This was the first blended club event (held in person and simultaneously online) since the pandemic started, and he noted that much preparation went into the event using no less than five laptops, a video camera, a mixing desk, many cables plus sound and projection systems (many thanks to Alex King and Brian Jepson for solving all the technology problems).

Risks on a Plane

John Spriggs, an independent writer and presenter, started the day with "*Risks on a Plane*"[1]. He gave an introduction to risk, its component parts and how it can be represented within a two-dimensional plane with severity and likelihood.

He had several references to its use in aviation and the software assurance guidelines, DO-178 and DO-278. He gave some definitions of risk from international standards and explained how there were many different definitions which were not consistent. Organisational risk appetites and risk matrices were covered. He noted that risk matrices need to be maintained and reviewed as things changes – he suggested every two years is about right.

Example Risk Matrix with Risk Classes

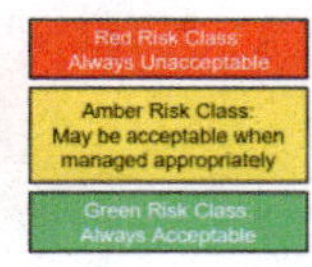

He explained the concept of Safety Objectives as invariants for that system and organisation. He summed up with "Establish your number system, then use it as the basis of your Risk Classification Scheme, which is developed by eliciting the client's risk appetite. Document it, declaring all assumptions". Addressing the topic of the seminar, he explained that to the question "Can we Quantify Risk?" the answer is "Yes, you can quantify risk but be careful how you use the numbers...".

Risk quantification with a lot of data (but limited knowledge) – Building a road risk tool after Selby

James Catmur of J C & A explained some of the situations he had been in over his career in rail and road. He gave some of the background to the Selby accident in 2001 when a vehicle crashed down an embankment and caused a train to derail and another train to crash into the wreckage. Ten lives were lost and 82 injured, the worst rail disaster of the 21st century in the UK [1].

He explained how road crash maps are maintained for UK (crashmap.co.uk) and lots of data on causes is available via the UK government website [2]. He said that human understanding of risk is very biased: people have a tendency to overestimate the frequency/risk of things they have experienced, underestimate the frequency/risk of things they have never experienced, believe they know all about road safety and use 'rationally motivated ignorance', i.e. "what you don't know can't hurt you".

[1] A nice pun given John's aviation background!

He said that risk assessments need to be kept simple and understandable, and the logic (i.e. the working or methodology) behind any numbers produced should always be shown. On the roads we have both Unsafe Acts and Unsafe Conditions so both need to be factored in, but they are different. When managing risks, it is important not to push risk from one group to another, i.e. it would be easier to make some roads lower risk for cars but higher risk for motorcycles. He said we need to seek to reduce our ignorance, i.e. find information to fill in the gaps. There was a lively debate about abuse of hard shoulder lanes on motorways and he explained that you should always wait outside the vehicle and upstream.

Quantitative Risk Analysis: Purpose, Prediction, Problems & Possibilities

John McDermid was on next after the coffee break and explained the difference between Retrodiction and Prediction with the former being the assessing of actual risk, posed by some (class of) system operating in an environment, over some period of time. He used the Boeing accident data plot to illustrate:

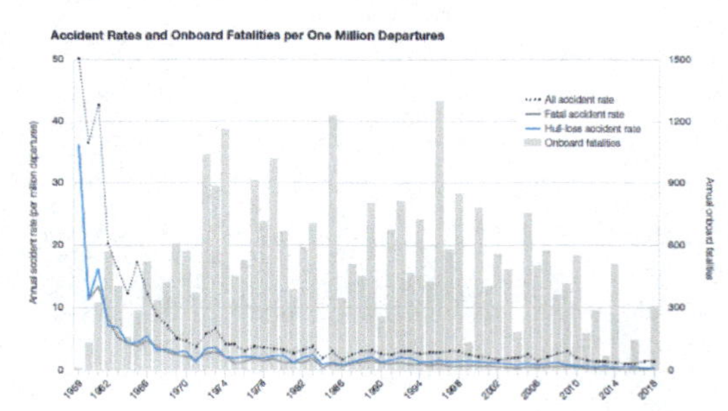

Prediction is where we estimate risk which will be posed by some (class of) system operating in an intended environment prior to deployment (or update), perhaps to inform regulatory approval or to inform insurance. He explained that we should be operating on a continuum – updating predictions from operational data, but this rarely happens.

John had some good quotes to illustrate that prediction is difficult: "Prediction is very difficult, especially if it's about the future."[2] and "Remember, John, if a safety case contains numbers, then they are wrong."[3]

He noted that the Watchkeeper accident rate appeared to much higher than initially predicted with several documented accidents. The majority of John's talk was taken up with an assessment framework and maturity model for Quantitative Risk Assessments based on a study and paper "Fixing the cracks in the crystal ball: A maturity model for quantitative risk assessment.

Fixing the cracks in the crystal ball: A maturity model for quantitative risk assessment

Andrew Rae, Rob Alexander*, John McDermid[1]

John listed some of the issues with these models: (i) if the system is safe enough then it may not get any meaningful feedback during the operational life, (ii) models normally assume stationary stochastic processes, but the environment and system change, and (iii) they don't account for "black swan" (very rare and very severe) events.

[2] Attributed to Niels Bohr (various versions). Also attributed to Mark Twain, Yogi Berra ...

[3] Former BAe Military Aircraft Chief Safety and Airworthiness Engineer.

He then outlined the maturity model with five levels: **Unrepeatable**, e.g. data sources not stated or analysis pre-dates the final design, **Invalid**, e.g. human/software causes not considered in accident sequences (incomplete, partial), **Valid but inaccurate**, e.g. incorrect assumptions on independence, **Accurate but challengeable**, e.g. use of data from other systems is controversial and **Ideal** – unattainable perfection.

He finished by considering assessment of AI/ML systems illustrated by a discussion of the Uber Tempe accident. His conclusion was that QRA can lead to false confidence in systems but it is possible to utilise the maturity model to get "better" QRA and a basis for reasoning about the figures in a safety case. He said the AAIP's main focus is on autonomy but noted that the performance measures for systems involving machine learning (e.g. for object recognition on the road) are typically in the range of 9X%, not 10^{-X} as we might hope.

Varieties of Risk

Peter Ladkin of Causalis started his talk with a short history of risk estimation and its origins in the insurance industry in London where ships and their cargos could be covered via negotiations conducted in London coffee houses:

He detailed the many and varied definitions of risk in international standards. He explained that there can be both positive and negative risks. The distinction of uncertainty and risk was historically defined as "Risk is when you know the probabilities; the die is fair. Uncertainty is when you don't know if the die is fair or not."

The Fukushima nuclear accident was discussed and the problems of insuring nuclear risks highlighted. In the UK, Nuclear Peril, i.e. environmental release of radioactive material or breach of the Reactor Pressure Vessel (RPV) is not insured commercially and is something for government. World-wide nuclear accidents were discussed and the possibility that Chernobyl could have been so much worse if a radioactive cloud had landed on Kiev.

Cyber-security risks were thought to be another sort of risk, as they are dynamic as opposed to safety risk which is generally considered static.

How the insurance industry quantifies and prices risk

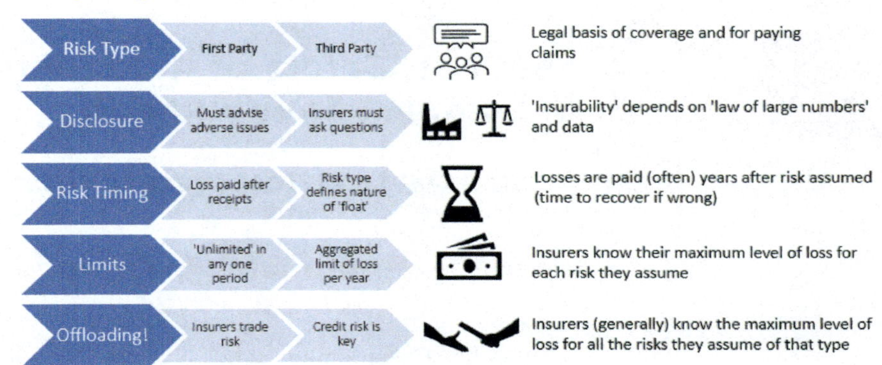

Clive Thompson of CTRL/2 gave a very interesting perspective of risk from the insurance industry. He explained that risk was viewed as opportunity for the industry, and it was a massive global business. Insurers need to be able to quantify risk to be able to calculate premiums. He explained five key aspects:

- **Risk Type**: first party e.g. buildings and contents vs. 3rd party such as public liability
- **Disclosure**: the underwriter needs information to assess the risk
- **Risk Timing**: including the importance of paying the premium in advance of any claim
- **Limits**: i.e. caps on payouts depending on risk taken on
- **Offloading**: spreading risks by using re-insurers

He explained that now in the digital age, multiple sources of information (e.g. big data analytics and satellite data for weather risks) are available to assist. Insuring some new types of risk e.g. autonomous road vehicles was also discussed.

The day finished with a lively panel discussion with four of the presenters (with three in the room and Peter Ladkin from Germany).

It was felt the day went very well with interesting and useful presentations, good discussions, and the technology for blended seminars had performed better than expected.

References

[1] https://en.wikipedia.org/wiki/Selby_rail_crash, accessed October 2021

[2] https://www.gov.uk/government/statistical-data-sets/reported-road-accidents-vehicles-and-casualties-tables-for-great-britain, accessed October 2021.

Image attribution
top image: 158612573 © Mitch Hutchinson | Dreamstime.com
risk game: Image of the game version produced by John Waddington Ltd. RISK! Is now owned by Hasbro Inc.
Selby crash: from The Guardian and Design Manual for Roads and Bridges
Fukushima: TEPCO CC BY-SA 2.0

Seminar: Safety of Autonomy in Complex Environments

THE SAFETY-CRITICAL SYSTEMS CLUB, Seminar:

Safety of Autonomy in Complex Environments

Thursday 22 September, 2022 - London, UK and blended online

This 1-day seminar will consider the safe use of autonomy in complex environments (for example a self-driving vehicle in a city environment), based on the work undertaken over the last couple of years at the Assuring Autonomy International Programme (AAIP) at the University of York. This work has produced a framework document, "Guidance on the Safety Assurance of Autonomous Systems in Complex Environments (SACE)" authored by Richard Hawkins, Matt Osborne, Mike Parsons, Mark Nicholson, John McDermid and Ibrahim Habli. This outlines techniques and approaches for assurance and gives an example safety argument.

Further details TBA.

Safe use of Multi-Core and Manycore Processors

This 100th Safety-Critical Systems Club (SCSC) seminar was held both face to face and online on 11th Nov 2021. The topics centred on approaches for using multi and many core processors (MCP) in safety-related and safety-critical applications. This is a new field for industry, and there are many challenges to produce a suitable assurance argument.

Mike Parsons introduced the seminar speakers and opened by noting that the use of such processors has been around for a while. However, the concept of using (and proving!) them in safety-critical applications is new.

Multi and Manycore Safety Working Group (MCWG)

Lee Jacques from Leonardo (and co-chair of the working group) gave an overview of the group covering past, present and future plans.

The past

Lee explained the (relatively young) history of the group and that it was created to discuss the challenges around multicore certification and the creation of CAST-32A – a multicore position paper by the Certification Authorities Software Team (CAST). The group has made some good progress in creating a common ontological model, and a number of sub groups have shared knowledge and information thus creating some strong networks.

Present

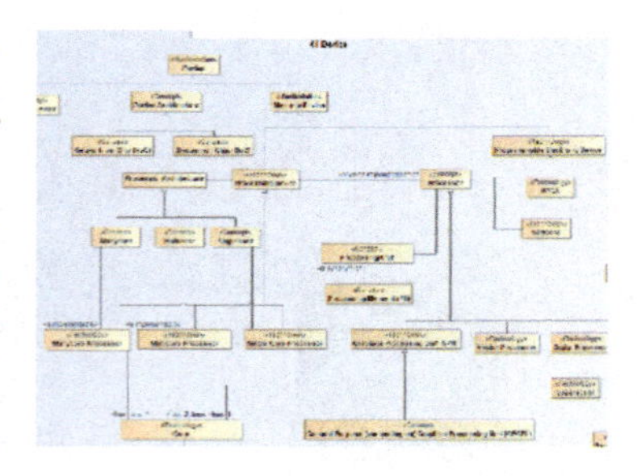

Whilst initially making good progress Lee highlighted that, the group has started to struggle, as the scope of the sub groups was too wide and introduced significant overlap. The ontology group however, was able to maintain a good pace as it has a defined and bounded scope of activities, which provided the team with a clear focus. Other challenges related to resource availability and the sharing of specific Intellectual Property (IP) were noted.

Future

Lee explained that the group was intending to refresh its approach and take a leaf from the ontology group be defining a clear set of well-defined tasks. The current approach was trying to cover a set of objectives that were too broad and a defined focus on clear smaller tasks would be more effective.

Lee asked the group for ideas and challenges that can be used to build a backlog of tasks, which the group can tackle in this new approach.

The Safety and Security Considerations for the Use of Multi-Core Processors

Mike Standish from DSTL and Mark Hadley from Atkins, presented how safety and security should be considered when undertaking certification of MCPs.

They noted a number of challenges including an inability to make an "in service" case, as this technology is not generally in use within a safety-critical component yet. Also, the complexity of these devices means that it is critical that the user understands the device, its architectural performance, boundary, longevity and specific configuration. Without understanding all of these parameters, the ability to make a reasoned assurance case is challenging.

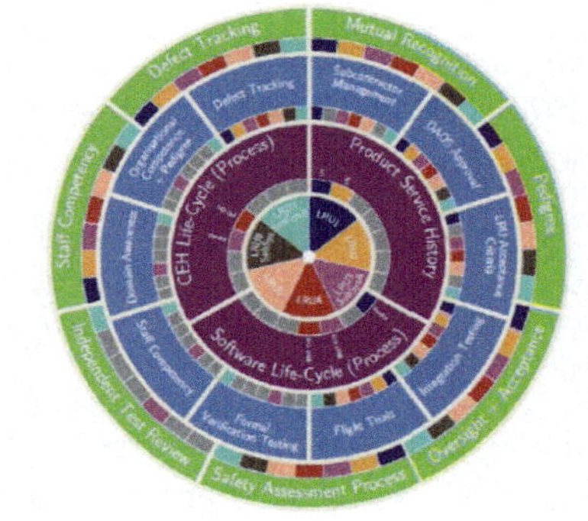

They introduced the "wheel of qualification" (Hadley & Standish 2019) and emphasised the need to make a diverse assurance case relying on various types of evidence and not just (for example) a measure of worst-case execution time (which although perfectly valid, cannot be used alone to justify assurance).

Although a challenge for MCP certification, Safety is relatively well-known in terms of process and assurance. Security on the other hand, introduces additional challenges and does not live in harmony with safety. For example, security patches, whilst necessary, can invalidate (or at least cause a re-evaluation) of an assurance case.

Mike and Mark's closing note was that MCP certification is not just about the MCP, it's about the whole ecosystem around the solution. This extends from the hardware setup and development environment through to the supply chain.

Incremental Assurance of Multicore Integrated Modular Avionics

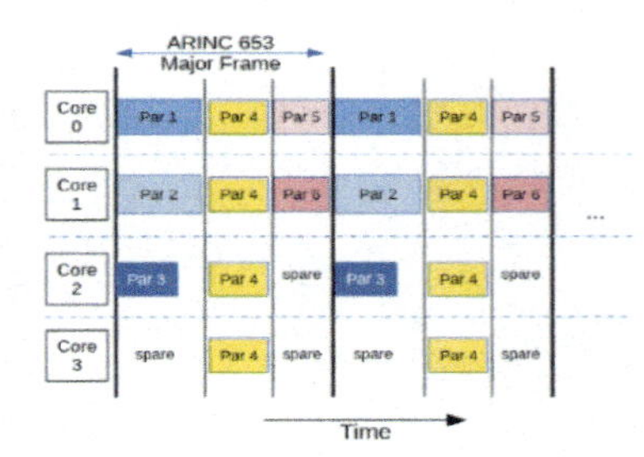

Guillam Bernat from Rapita, discussed the challenges of certifying Integrated Modular Avionics (IMA) with MCP solutions and techniques for performing performance analysis. Using Rapita tools, it is possible to monitor all aspects of MCP performance using embedded RapiDaemons. The challenge with a traditional IMA solution means a significant amount of recertification is required when changing one item.

Guillam stated that careful consideration of the partitioning model and use of automated testing is critical to success. Whilst it is possible to automate data gathering activities, there is a significant amount of manual analysis required to interpret the data.

The presentation highlighted potential test solutions using interference generators to mitigate the challenge of identifying and verifying interference paths in a multicore solution. Guillam continued detailing how this could be used in a mixed criticality context, crucial for keeping time and costs down and easing the certification burden.

Multicore Processors usage in Certified Avionics: How Virtualisation Can Help?

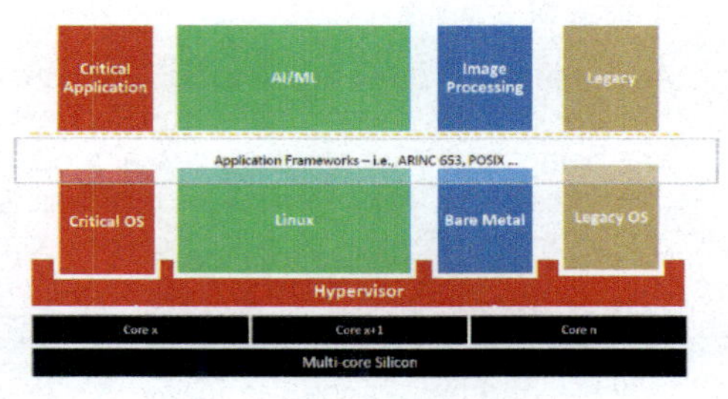

Olivier Charrier from WindRiver, explained how the use of virtual machines can provide the assurance required when partitioning multicore systems. Olivier stressed the importance of considering the architecture and requirements early, and just as importantly as the test strategy. He highlighted that as important as this is, it's also important to consider that assumptions made early on in the process may not fully hold going forward, and that continuous test and evaluation can drive design choices.

The presentation then went on to focus on the potential benefits of using virtualisation to provide a complete partitioning solution (memory, CPU, Cache, etc). Using virtualisation (managed by a Hypervisor) would support an assurance case by providing a well partitioned solution and would address many of the resource usage aspects of CAST-32A. That is of course, assuming you have done the upfront work to determine that virtualisation is an appropriate architecture for your solution!

Telemetry and bare-metal Virtual Machines for Improved Multicore Partitioning

Tim Loveless from Lynx Software Systems gave an overview of Hypervisor technology, which was widely regarded in the seminar as one of the clearest definitions people had seen.

Tim started by noting that most people assume that to ensure a well-partitioned system, you need an ARINC653 based Real-Time Operating System (RTOS). Whilst in many cases this is the correct solution, he noted that for some architectures it is possible to run a bare metal hypervisor to improve performance and reduce complexity'. Do you really need that Ethernet stack and file system?

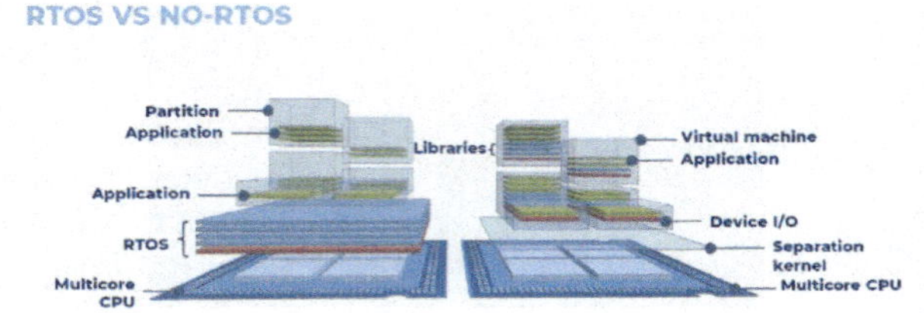

Tim introduced a number of potential patterns, which could be used to deliver a compliant bare metal solution and also introduced the CPU Performance Management Unit (PMU). This is a key component when testing and characterising your CPU as it provides a series of counters measuring everything from number of instructions completed to data misses. He warned though although important for characterisation, they introduce an overhead into the system and the more complex the system, the more integration with the RTOS is required. Think how often, how and when you are going to store or offboard the data whilst trying to maintain multicore, real-time performance...

Multi-core architectures and timing analysis: Their influence on the scheduling of certifiable real-time systems

Iain Bate from the University of York presented some techniques for analysing multicore architectures and building a timing analysis approach to most effectively assess the performance of a system.

As many other presenters noted, it is key that all of this activity needs to be considered up front and that system architecture understanding is key. For example, what CPU resources are being used, what resources are being shared, what partitioning approach should be considered etc. Making design decisions is difficult as the software implementation affects the performance, but at this stage the software has not been written.

Iain presented a 4-step process to introduce some rigour into the process and help guide the identification and mitigation of the key interference factors.

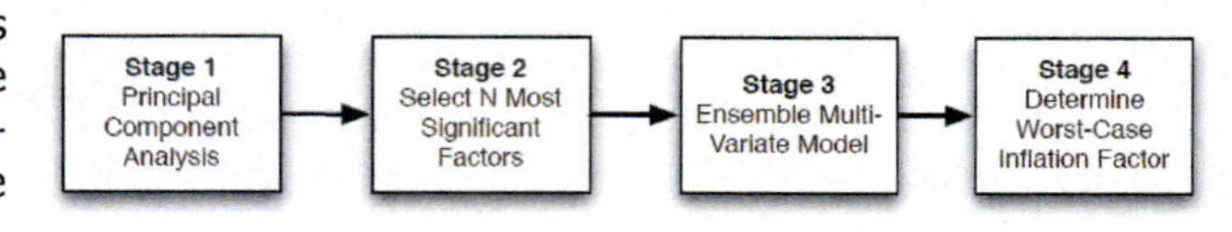

Certification aspects of Multicore

Sam Riley from Frazer Nash (Formerly MAA) gave an insight into the thinking of a certification authority based on first-hand experience and noted that there is no definitive approach and positive engagement with the authority is key.

He concluded by discussing 7 general "lessons" from his experience to help those preparing safety cases, for example, ensuring that evidence is diverse and that key people, such as the design leads and regulator, are taken "along on the journey" from an early stage.

Report by Lee Jacques, MCWG Co-Chair

Getting to Know You An update from the Safety Futures Initiative

As we've seen from our previous forward-looking articles, developing the next generation of safety engineers will be critical to ensuring the club can continue to make systems safer over the next 30 years. Zoe Garstang, lead for the Safety Futures Initiative (SFI), provides an update on the progress made by the SFI and provides details of future events.

Get To Know You Event

The Safety Futures Initiative (SFI) held their second set of 'Get to Know You Events' on 24th November 2021, with a lunchtime and evening session. The presentation material used at these sessions is available on the SFI webpage (www.scsc/gf).

The events built on the feedback received from the July events and planning is underway for new activities throughout 2022, including a lecture competition.

Due to the changing nature of these activities, please keep checking the SFI webpage for the most recent updates.

Looking Ahead

There will be further regular Get to Know You Events for new and existing members to come and find out more about the group and its upcoming activities. The dates for 2022 will be communicated via the website, social media and directly to SFI members.

At the Safety-Critical Systems Symposium (SSS'22), taking place on 8-10th February 2022 (https://scsc.uk/e797), there will be an opportunity to learn more about the SFI and meet new and existing members in person.

This will take place as part of the Working Group's session at 9:00am-10:00am on Wednesday 9th February 2022.

"planning is underway for new activities throughout 2022, including a lecture competition"

Membership

The first year's membership of the SFI is free, so I would encourage anyone who would like to get involved to sign-up (please see www.scsc.uk/membership).

SFI members get access to all SFI events and activities, as well as discounted fees at SCSC Events.

Further Information

If you are unable to attend SSS'22 or would like further information about the SFI, please do get in touch with Zoe Garstang (zoe.garstang@scsc.uk).

Zoe Garstang, Airworthiness Engineer and SFI Lead

Zoe is a Flight Safety Analyst at BAE Systems, providing in-service support to the Typhoon aircraft. She previously undertook an Advanced Engineering Apprenticeship with the company before joining the Continued Airworthiness team.

Connect

The Newsletter and eJournal

Do you have a topic you'd like to share with the systems safety community? Perhaps an interesting area of research or project work you've been involved in, some new developments you'd like to share, or perhaps you would simply like to express your views and opinions of current issues and events. There are now two publishing vehicles for content – shorter, more informal content, can be published in the Newsletter with longer, more technical peer-reviewed material more suitable for the eJournal. If you are interested in submitting content, then get in touch with Paul Hampton for Newsletter articles: paul.hampton@scsc.uk or John Spriggs for eJournal papers: john.spriggs@scsc.uk

Authors of papers published in this Newsletter or in the eJournal will be offered a year's free membership of the Safety-Critical Systems Club.

The SCSC Website

Visit the Club's website thescsc.org for more details of the Safety-Critical Systems Club including past newsletters, details of how to get involved in working groups and joining information for the various forthcoming events.

Facebook

Follow the Safety-Critical Systems Club on its very own Facebook page.

www.facebook.com/SafetyClubUK

Twitter

Follow the Safety-Critical Systems Club's Twitter feed for brief updates on the club and events: @SafetyClubUK

LinkedIn

You can find the club on LinkedIn. Search for the Safety-Critical Systems Club or use the following link:

www.linkedin.com/groups/3752227

Advertising

Do you have a product, service or event you would like to advertise in the Newsletter? The SCSC Newsletter can reach out to over 1,000 members involved in Systems Safety and so is the perfect medium for engaging with the community. For prices and further details, please get in touch with the Newsletter Editor.

SCSC Working Groups

The Safety-Critical Systems Club is committed to supporting the activities of working groups for areas of special interest to club members. The purpose of these groups is to share industry best practice, establish suitable work and research programmes, develop industry guidance documents and influence the development of standards.

Assurance Cases

The Assurance Cases Working Group (ACWG) has been established to provide guidance on all aspects of assurance cases including construction, review and maintenance. The ACWG will:

- Be broader than safety, and will address interaction and conflict between related topics
- Address aspects such as proportionality, rationale behind the guidance, focus on risk, confidence and conformance
- Consider the role of the counter-argument and evidence and the treatment of potential bias in arguments

In Aug 2021, the group published v1.0 of the Assurance Case Guidance: scsc.uk/scsc-159

One of the working group's activities is the maintenance of the Goal Structuring Notation (GSN) Community standard.

See scsc.uk/gsn for further details.

In May 2021, the group published v3.0 of the standard: scsc.uk/scsc-141C

Lead **Phil Williams** phil.williams@scsc.uk

SCSC Working Groups

Security Informed Safety

The Security Informed Safety Working Group (SISWG) aims to capture cross-domain best practice to help engineers find the 'wood through the trees' with all the different security standards, their implication and integration with safety design principles to aid the design and protection of secure safety-critical systems and systems with a safety implication.

The working group aims to produce clear and current guidance on methods to design and protect safety-related and safety-critical systems in a way that reflects prevailing and emerging best practice.

The guidance will allow safety, security and other stakeholders to navigate the different security standards, understand their applicability and their integration with safety principles, and ultimately aid the design and protection of secure safety-related and safety-critical systems.

Lead **Stephen Bull** stephen.bull@scsc.uk

Data Safety Initiative

Data in safety-related systems is not sufficiently addressed in current safety management practices and standards.

It is acknowledged that data has been a contributing factor in several incidents and accidents to date, including events related to the handling of Covid-19 data. There are clear business and societal benefits, in terms of reduced harm, reduced commercial liabilities and improved business efficiencies, in investigating and addressing outstanding challenges related to safety of data.

The Data Safety Initiative Working Group (DSIWG) aims to have clear guidance on how data (as distinct from the software and hardware) should be managed in a safety-related context, which will reflect emerging best practice.

An update to the guidance (v3.4) was published in Jan 2022: scsc.uk/scsc-127G

Lead **Mike Parsons** mike.parsons@scsc.uk

SCSC Working Groups

Safety of Autonomous Systems

The specific safety challenges of autonomous systems and the technologies that enable autonomy are not adequately addressed by current safety management practices and standards.

It is clear that autonomous systems can introduce many new paths to accidents, and that autonomous system technologies may not be practical to analyse adequately using accepted current practice. Whilst there are differences in detail, and standards, between domains many of the underlying challenges appear similar and it is likely that common approaches to core problems will prove possible.

The Safety of Autonomous Systems Working Group (SASWG) aims to produce clear guidance on how autonomous systems and autonomy technologies should be managed in a safety-related context, in a way that reflects emerging best practice.

The group published v3 of its guidance Safety Assurance Objectives for Autonomous Systems, in Jan 2022 scsc.uk/scsc-153B

Lead Philippa Ryan pmrc@adelard.com

Multi- and Manycore Safety

It is becoming harder and harder to source single-core devices and there is a growing need for increased processing capability with a smaller physical footprint in all applications. Devices with multiple cores can perform many processes at once, meaning it is difficult to establish (with sufficient evidence) whether or not these processes can be relied upon for safety-related purposes.

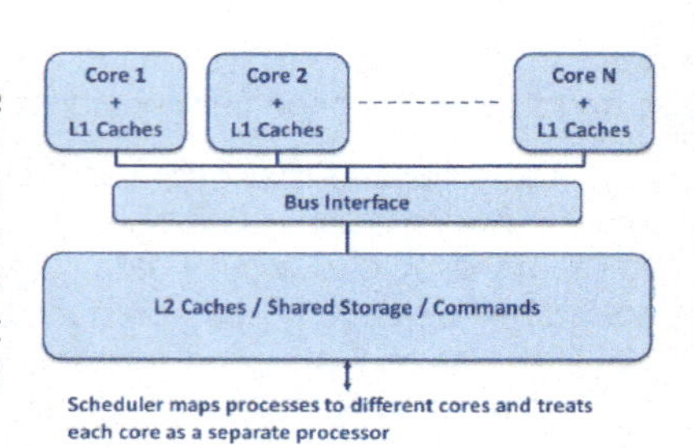

Scheduler maps processes to different cores and treats each core as a separate processor

Parallel processes need to access the same shared resources, including memory, cache and external interfaces, so they may contend for the same resources. Resource contention is a source of interference which can prevent or disrupt completion of the processes, meaning it is difficult to know with a defined uncertainty the maximum time each process will take to complete (Worst Case Execution Time, WCET) or whether the data stored in shared memory has been altered by other processes.

The Multi- and Manycore Safety Working Group (MCWG) has been established to explore the future ways of assuring the safety of multi- and manycore implementations.

Lead Lee Jacques Lee.Jacques@leonardocompany.com

SCSC Working Groups

Ontology

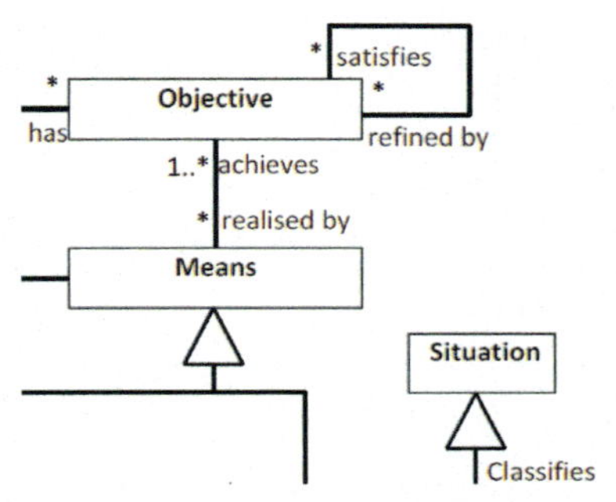

The Ontology Working Group (OWG) develops ontologies that will form the basis of SCSC guidance, as well as having wider industrial and academic applications.

The OWG is currently working on the definition of an ontology of risk for application in guidance for risk-based decision making – notably safety and security – and for which ISO 31000 Risk Management principles are to be applied.

The Data Safety Working Group (DSIWG) developed the core aspects of the Risk Ontology, which has been migrated to this working group. The Risk Ontology will form the upper ontology to the Data Safety Ontology that the DSIWG will continue to develop.

Lead Dave Banham ontology@scsc.uk

Covid-19

The Covid-19 Working Group is involved with discussion, analysis and assistance related to the Coronavirus. The group meets remotely to see what a systems and assurance view of the situation brings.

The group has compiled an extensive range of Covid-19 related material and made this available on the working group's website pages along with ongoing developments in the thoughts and ideas of the group.

Members are all experienced engineers, used to making reasoned arguments about safety. The aim is to apply the groups considerable technical expertise to the problem and find and assure appropriate solutions.

Lead Peter Ladkin ladkin@causalis.com

SCSC Working Groups

Service Assurance

Risks presented by safety-related services are rarely explicitly recognised or addressed in current safety management practices, guidelines and standards. It is likely that service (as distinct from system) failures have led to safety incidents and accidents, but this has not always been recognised. The Service Assurance Working Group (SAWG) has been set up to produce clear and practical guidance on how services should be managed in a safety-related context, to reflect emerging best practice.

The group published v3.0 of the guidance in Jan 2022: scsc.uk/scsc-156B

Lead Mike Parsons mike.parsons@scsc.uk

SCSC Safety Culture

The Safety Culture Working Group (SCWG) has been established to provide guidance on creating and maintaining an effective safety culture. The group seeks to improve safety culture in safety-critical organisations focussed on product and functional safety, by sharing examples and latest approaches collated from real-life case studies.

Meetings provide an opportunity to discuss any particular aspects attendees are interested in taking forward, and to help set future directions for the group.

The group is planning to hold an event early in 2022.

Lead Michael Wright michael.wright@greenstreet.co.uk

60 Seconds with ... Dr Mike Parsons

Mike has worked for many years in the safety industry across diverse sectors such as Defence, Aerospace, Rail, Healthcare, Nuclear and Government. He has worked for Logica, CGI, NATS, and he is now a research fellow in the AAIP at the University of York working on the assurance of autonomy.

His career started back in 1988 when he worked on medical imaging systems, progressing through projects including space launch tracking systems, satellite navigation, healthcare applications and civil aviation messaging.

He is also the SCSC Director and Events Organiser, and leads two of the SCSC Working Groups on Data Safety and Service Assurance.

You've said before that your childhood dream was to be an astronaut. What aspect of space travel interests you the most?

I think it's the idea of complex systems working in a difficult and unforgiving environment and enabling new discoveries and applications. The idea that the systems in space stations, Mars rovers or deep space probes have to be super-resilient and be able to work autonomously is fascinating to me.

What aspect of your career are you most proud of?

That's easy: building communities of safety engineers to achieve a goal. I did this working at Logica where I created the safety community and also a safety practice; but to me, the SCSC embodies this completely. Becoming Director of the SCSC and running the Safety-Critical Symposium gives me a real buzz! I am really proud of the work I have done on the SCSC Working Groups (Data Safety and Service Assurance) and also at events where we all have a strong common purpose and work together to achieve it. In terms of projects, it would have to be the Ariane launch vehicle tracking and monitoring system I did for Logica and subsequently installing the kit at the launch site in French Guiana.

What advice would you give to yourself age 12?

Follow your dreams, but be prepared to be side-tracked! I think my dream of being an astronaut led me to work in the Space Sector, then in Safety, then for the SCSC. Safety is such an interesting and challenging area: I like the way it requires systems thinking as well as an appreciation of wider things like legal and ethical aspects, together with the constraints of what is possible. I would never have appreciated this as a 12-year-old, but it's important to give your career time to explore: some roles you don't even know exist may suddenly appear. I would also say don't be afraid: reach out and take on something new – you will find a way – and opportunities don't appear twice.

What worries you the most about the future of System Safety?

I am really concerned that our techniques for analysis of complex systems are not up to the job; systems are becoming ever more dynamic and distributed, with vast hidden complexities (and autonomous functionalities). Also, security needs to be properly integrated with safety. I think we need a whole new suite of powerful tools allowing us to reason about these systems. Historically, safety engineering has learnt from accidents. What worries me is that we'll have some terrible accidents (e.g. involving self-driving vehicles) and not be able to work out why it happened...

What's your most favourite quote or motto?

I always liked a saying we used to have at Logica when reviewing risks on new projects "Would it pass the headline test?" In other words, if the system or software developed went wrong and an accident resulted, could you construct a snappy 'tabloid-style' headline blaming us? If you can, then the system is likely not safety engineered enough...

If you could learn to do anything, what would it be?

Learn how to move around in a zero-g environment. Since being an astronaut is still somewhat unlikely, this might have to be done in one of the commercial zero-g flights now on offer. Also I think it would great to dive to the deepest ocean floor in a submarine – anything which takes me to new environments.

If you could be any fictional character, who would you choose?

Possibly Mark Watney from The Martian – I love the way he solved the hard problems while abandoned on Mars. I do like a good detective story, and I think Max Liebermann from Vienna Blood is an interesting twist.

"Becoming Director of the SCSC and running the Safety-Critical Symposium gives me a real buzz!"

What's the best piece of advice you've ever been given?

Perversely, "take a risk!" This was explained to me back in 1995 on my first Logica project and is important – safety engineering is all about managing risk, not eliminating it altogether.

Which song title best sums up your experiences with Covid-19?

"*David Sylvian – World Citizen (I Won't Be Disappointed) / Looped Piano Version*" sums up the sense of alienation, isolation and strangeness of the pandemic with a safety theme. I also really like *"Patricia Barber - Icarus (For Nina Simone)"* - I've listened to a lot more jazz over the last two years and this is a lovely example, with a risk and aviation topic.

SCSC Membership

The SCSC provides a range of services to the System Safety community including seminars, tutorials, leadership events, specialist topic working groups, the annual symposium and a comprehensive body of publications. Membership brings many valuable benefits such as free access to online events, the SCSC Newsletter and access to presentations and other resources from events.

Individual Membership

To become an individual member of the SCSC please register on the SCSC website using the icon at the top right of any page and select "Register". Complete and save your account registration and then verify your email address. Once registered and logged in click the link "why not join the SCSC..." inviting you to become a member at the top right of the page or select "Pay membership" from the icon.

Individual membership can be paid online using a credit/debit card through our secure payment partner Realex Global Payments or contact Alex King for other payment methods. For student or retired member rates please contact Alex King to get your account status changed.

Corporate Membership

Your company contact with the SCSC should arrange the membership and any renewals for your organisation. To join as a member covered by a corporate membership, register as per the instructions for an individual member and then contact Alex King to confirm your affiliation.

Renewing Membership

You should be notified by email when your membership is almost expired or shortly after it has expired. These notifications will contain a link to the online renewal page or you will be able to renew when logging onto the website through the 'click to renew' link.

Membership Fees

The following fees are applicable for new and renewing members:

- 1 year Individual Membership: £125
- 2 year Membership: 20% discount: £200
- 3 year Membership: 33% discount: £250 (3 years for the price of 2)
- 1 year SFI Membership: FREE for first year, £35 for years 2 & 3
- 1 year Membership, retired member rate: £35
- For Corporate Membership discounts contact Alex King.

A one-month Publication Pass is also available for £15. This allows access to all SCSC publications in a particular calendar month.

Contact Alex King using office@scsc.uk

The SCSC Steering Group

Tom Anderson

Honorary member

Stephen Bull

stephen.bull@scsc.uk

Jane Fenn

jane.fenn@scsc.uk

Paul Hampton

paul.hampton@scsc.uk

James Inge

james.inge@scsc.uk

Graham Jolliffe

Honorary member

Alex King

alex.king@scsc.uk

Wendy Owen

wendy.owen@scsc.uk

Felix Redmill

Honorary member

John Spriggs

john.spriggs@scsc.uk

Phil Williams

phil.williams@scsc.uk

Robin Bloomfield

Honorary member

Dewi Daniels

dewi.daniels@scsc.uk

Zoe Garstang

zoe.garstang@scsc.uk

Louise Harney

louise.harney@scsc.uk

Brian Jepson

brian.jepson@scsc.uk

Tim Kelly

Honorary member

Mark Nicholson

mark.nicholson@scsc.uk

Mike Parsons

mike.parsons@scsc.uk

Roger Rivett

roger.rivett@scsc.uk

Emma Taylor

Honorary member

Sean White

sean.white@scsc.uk

Club Positions

The current and previous (marked in italics) holders of club positions are as follows:

Managing Director

Mike Parsons 2019-

Tim Kelly 2016-2019

Tom Anderson 1991-2016

Steering Group Chair

Roger Rivett 2019-

Graham Jolliffe 2014-2019

Brian Jepson 2007-2014

Bob Malcolm 1991-2007

Programme & Events Coordinator

Mike Parsons 2014-

Chris Dale 2008-2014

Felix Redmill 1991-2008

Manager

Alex King 2019-

Newsletter Editor

Paul Hampton 2019-

Katrina Attwood 2016-2019

Felix Redmill 1991-2016

University of York Coordinator

Mark Nicholson 2019-

eJournal Editor

John Spriggs 2021-

Administrator

Alex King 2016-

Joan Atkinson 1991-2016

Website Editor

Brian Jepson 2004-

Safety Futures Initiative Lead

Zoe Garstang 2019-

Nikita Johnson 2019-2021

Calendar

February '22

M	T	W	T	F	S	S
	1	2	3	4	5	6
7	8	9	10	11	12	13
14	15	16	17	18	19	20
21	22	23	24	25	26	27
28						

March '22

M	T	W	T	F	S	S
	1	2	3	4	5	6
7	8	9	10	11	12	13
14	15	16	17	18	19	20
21	22	23	24	25	26	27
28	29	30	31			

April '22

M	T	W	T	F	S	S
				1	2	3
4	5	6	7	8	9	10
11	12	13	14	15	16	17
18	19	20	21	22	23	24
25	26	27	28	29	30	

May '22

M	T	W	T	F	S	S
						1
2	3	4	5	6	7	8
9	10	11	12	13	14	15
16	17	18	19	20	21	22
23	24	25	26	27	28	29
30	31					

June '22

M	T	W	T	F	S	S
		1	2	3	4	5
6	7	8	9	10	11	12
13	14	15	16	17	18	19
20	21	22	23	24	25	26
27	28	29	30			

July '22

M	T	W	T	F	S	S
				1	2	3
4	5	6	7	8	9	10
11	12	13	14	15	16	17
18	19	20	21	22	23	24
25	26	27	28	29	30	31

August '22

M	T	W	T	F	S	S
1	2	3	4	5	6	7
8	9	10	11	12	13	14
15	16	17	18	19	20	21
22	23	24	25	26	27	28
29	30	31				

September '22

M	T	W	T	F	S	S
			1	2	3	4
5	6	7	8	9	10	11
12	13	14	15	16	17	18
19	20	21	22	23	24	25
26	27	28	29	30		

October '22

M	T	W	T	F	S	S
					1	2
3	4	5	6	7	8	9
10	11	12	13	14	15	16
17	18	19	20	21	22	23
24	25	26	27	28	29	30
31						

November '22

M	T	W	T	F	S	S
	1	2	3	4	5	6
7	8	9	10	11	12	13
14	15	16	17	18	19	20
21	22	23	24	25	26	27
28	29	30				

December '22

M	T	W	T	F	S	S
			1	2	3	4
5	6	7	8	9	10	11
12	13	14	15	16	17	18
19	20	21	22	23	24	25
26	27	28	29	30	31	

January '23

M	T	W	T	F	S	S
						1
2	3	4	5	6	7	8
9	10	11	12	13	14	15
16	17	18	19	20	21	22
23	24	25	26	27	28	29
30	31					

Events Diary

8-10 February 2022 SCSC Symposium **30th Safety-Critical Systems Symposium (SSS'22)** **Bristol, UK + Online** scsc.uk/e797	**28-29 March 2022** **Conference** **16th International Conference on Safety and Systems Engineering (ICSSE 2022)** **Paris, France** waset.org/safety-and-systems-engineering-conference-in-march-2022-in-paris	**8 April 2022** SCSC Seminar **Managing 'Black Swans': Handling Rare and Severe Events Now and in the Future** **London, UK + Online** scsc.uk/e825	**1-2 June 2022** Conference **Reliability, Safety and Security of Railway Systems (RSSRail 2022)** **Paris, France** rssrail2022.univ-gustave-eiffel.fr
6-9 September 2022 Conference **41st International Conference on Computer Safety, Reliability and Security (SAFECOMP 2022)** **Munich, Germany** safecomp22.iks.fraunhofer.de	**22 September 2022** SCSC Seminar **Seminar: Safety of Autonomy in Complex Environments** **London, UK + Online** scsc.uk/e890		

NB: all events are subject to change due to the Covid-19 situation. Please check the SCSC website for up-to-date information: scsc.uk/events

thescsc.org/membership

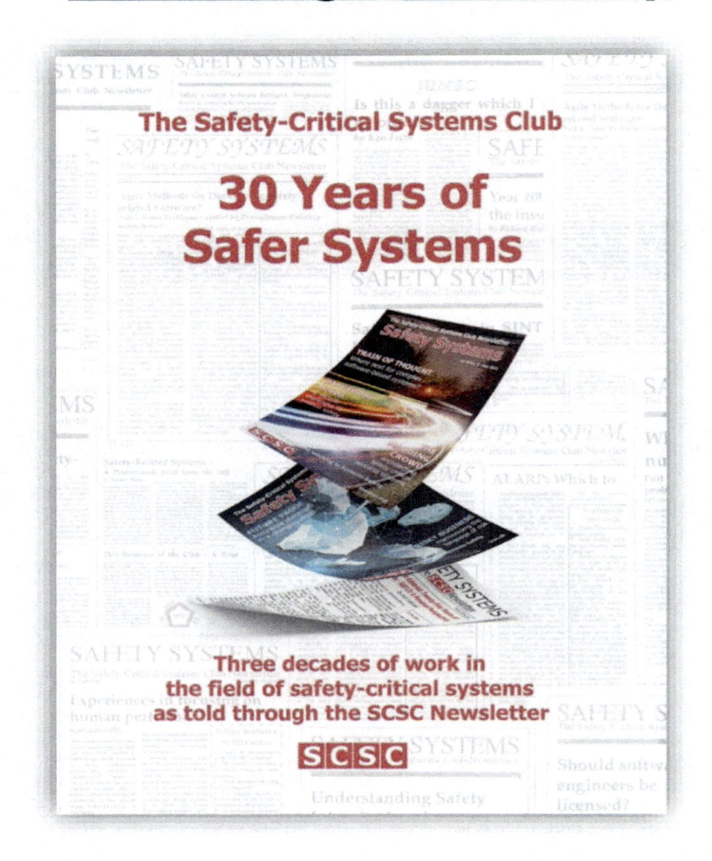

"30 Years of Safer Systems" contains articles from the last 3 decades of the Safety-Critical Systems Club (SCSC) newsletter "Safety Systems".

The book groups the articles into themes relevant to safety, with an introduction to the theme and a preface to each article giving major events from the year the article was first published, including accidents, incidents and positive improvements in safety. Themes include: Risk Assessment, ALARP, Artificial Intelligence/Machine Learning, Communication Failures, Safety Culture, 'Black Swan' events, Certification, Product Liability, Safety and Security Integration, Agile Methods, Data driven systems and Safety Cases.

Most of the original authors have provided a short postscript to their article to give extra context and explain progress in the intervening years.

Available for purchase on Amazon
www.amazon.co.uk/Years-Safer-Systems-safety-critical-Newsletter/dp/B09KNCYKDL

The Safety-Critical Systems Club Newsletter

Safety Systems

Vol 30 No. 2 - May 2022

COLLISION COURSE!

The road to safer autonomous vehicles

EXIT STRATEGY

Software liability in a post Brexit world

MEETING OF MINDS

Highlights from the SSS'22 Symposium

For everyone working in Systems Safety

thescsc.org

SCSC Publication Number: SCSC-175

Contents

www.scsc.uk/sss

SSS' 23

SCSC 31st Annual Symposium

Details of this and other events at: www.scsc.uk

www.scsc.uk

THE SAFETY-CRITICAL SYSTEMS CLUB

31st Safety-Critical Systems Symposium

7-9th February 2023, in-person in York and blended online

Invitation to submit an abstract

The Safety-Critical Systems Symposium in 2023 (SSS'23) will be held in York, UK and blended online as a live-streamed event. This event comprises three days of live presented papers, including keynote presentations and submitted papers.

The Symposium is for all of those in the field of systems safety, including engineers, managers, consultants, students, researchers and regulators. It offers wide-ranging coverage of current safety topics, focussed on industrial experience. It includes recent developments in the field and progress reports from the SCSC Working Groups. It covers all safety-related sectors including aerospace, defence, energy, healthcare, highways, marine, nuclear and rail.

Suggested topics are:

Accident Analysis
Agile Methods
Artificial Intelligence
Argument Notations
Assurance Cases
Autonomous Systems
CAV
Certification
Change and Evolution
Complex Systems & Environments
Digital Twins
EMC/EMI
Human Factors
Internet of Things
Machine Learning
Methods and Tools
Model-Based Assurance
Modular Assurance
Multicore / Manycore
Ontologies / Formalisms
Regulation of Safety
Resilience
Robotics and Automation
Safety Analyses
Safety Culture
Safety Management
Safety Practice and Process
Service-Oriented Safety
Software
Standards and Guidance
Systems-of-Systems
Through-Life Safety
Training and Training Data
Uncertainty
Validation and Verification

Authors for papers should submit a title and 250-word abstract to **mike.parsons@scsc.uk** by **30th June 2022**. Successful authors will then be asked to submit their paper by **30th October 2022**. Papers will be reviewed, and comments fed back during November 2022.

The proceedings book will be available at the event. For **information, exhibition** and **booking** enquiries please contact: Alex King, Dept of Computer Science, University of York, Deramore Lane, York, YO10 5GH. Tel: 01904 325402; **alex.king@scsc.uk.** For technical aspects, abstract and paper submissions, please contact: **mike.parsons@scsc.uk**

Editorial

With life largely getting back to normal in most parts of the UK, there does seem a perception that Covid-19 is behind us, or at least, there are other more pressing concerns to occupy our thoughts such as increasing fuel and living costs and, of course, the situation in Ukraine. While it is debatable whether Covid-19 is completely over, at least during the last two years we felt we could make a difference and contributed efforts in applying systems thinking to the problem, especially through the good work of the SCSC Covid-19 Working Group.

The current crisis in Ukraine is clearly having a global impact and represents our next biggest challenge; but in the chaos of war, it seems harder to apply systems thinking when the parties involved have such conflicting objectives. As reported in the news, the IAEA said 'nuclear safety has been seriously jeopardised' and one wonders how any safety case meets its claims in war time.

A serious concern arising from heightened political tensions is the accidental use of weapons, which could escalate into full-scale war, and these sorts of concerns are really where formal approaches to system safety began in the 1960's with the Minuteman Ballistic Missiles. I'd recommend reading the SCSC Newsletter article by John Ridgway 'On the Safest Way to Kill' (Volume 27 No. 1); John concluded that accidental detonation still presents the greatest existential threat to mankind. The recent accidental launch of a missile between India and Pakistan due to a "technical malfunction" is particularly sobering.

The SCSC celebrated its 30th anniversary and the annual Symposium held in February reflected on the huge achievements and contributions the club has made over that time. It also looked to the future and the challenges that lie ahead, and the need for systems thinking and safety engineering has never been more important. Autonomous Vehicles represents one of the major future challenges and in our first article, Roger Rivett discusses the challenges of self-driving cars and proposes ways to progress toward making this technology a reality on our roads. As well as technological disrupters, there have been some major changes politically in the exit of the UK from the European Union. Dai Davis discusses some of the implications that this has had on product liability.

There then follows a number of event reports starting with a report of the SSS'22 symposium, including summaries of all the key note speeches, and I have pleasure in being able to provide the transcripts from Tim Kelly's 'thought for the symposium' that concluded the event and Wendy Owen's pre-dinner talk. Michael Wright provides a report of the seminar 'Accident Investigation and Safety Culture' run by the Safety Culture Working Group. There is also a report from Dewi Daniels on the club's very first Technical Trip to Bletchley Park. This was a day trip, open to all, to the once top-secret home of the World War Two Codebreakers and is intended to be the first of many such visits to interesting locations like this.

Our 60 second interview is with Zoe Garstang – SCSC Steering Group member and lead for the Safety Futures Initiative.

Paul Hampton
SCSC Newsletter Editor
paul.hampton@scsc.uk

In Brief

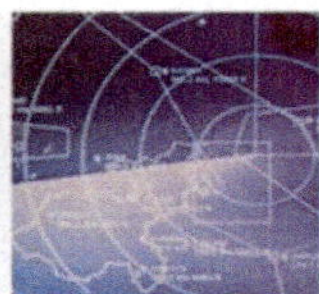

India accidentally fires missile into Pakistan

India says it accidentally fired a missile into Pakistan, blaming the incident on a "technical malfunction" during routine maintenance. Delhi said it was "deeply regrettable" and expressed relief no one was killed.

Pakistan's military said a "high-speed flying object" had crashed near the eastern city of Mian Channu and its flight path had endangered passenger flights. *bbc.co.uk*

Nuclear Safety in Ukraine 'Seriously Jeopardized on Several Occasions' – IAEA

Nuclear safety in Ukraine has been "seriously jeopardized on several occasions" since the start of the Russian invasion, according to Rafael Mariano Grossi, the director-general of the International Atomic Energy Agency (IAEA).

The IAEA team were in Chornobyl starting on 26th April, and delivered vital equipment that Ukrainian officials say the country needs for the safe and secure operation of its nuclear facilities. *newsweek.com*

Head of Hokkaido tour boat operator: Fatal accident could have been avoided

The president of a tour boat operator in Hokkaido Prefecture, northern Japan, has stated in a document that a fatal accident last month could have been avoided if safety protocols were followed. *www3.nhk.or.jp*

Pilot and Instructor Faulted in Crash that Destroyed $64M Global Hawk

A drone pilot and an instructor at Grand Forks Air Force Base, N.D., failed to realise an RQ-4 was 4,000 feet too high as it started its final approach and didn't select the proper commands, leading the unmanned aerial system to overshoot the base and crash nearly seven miles away, according to a new accident investigation board report. *uasvision.com*

Tesla recalls second batch of cars in China on safety concerns

US car giant Tesla has recalled more than 14,600 cars in China over a software fault that could lead to safety risks, the country's market regulator said on 29th April 2022, the firm's second callback in a month. *todayonline.com*

Autonomous Driving – more than just avoiding collisions

Roger Rivett discusses the challenges of having a fully autonomous vehicle using the current road infrastructure alongside non-autonomous vehicles and other road users. He examines whether changes could be made to the road infrastructure and extending the scope of the safety arguments, so that these are both more accommodating to autonomous vehicle technologies.

This article describes the motivation behind the creation of an ontological model of the public road transport system under the auspices of the Assuring Autonomy Internal Programme (AAIP), [1] and published on the AAIP website, [2]. That work results in a set of vehicle movement constraints that are being used in work being undertaken by the MISRA Safety Argument working group, [3].

There seems to be a general consensus among politicians and car manufacturers that the progressive introduction of Advanced Driver Assistance Systems (ADAS), [4], [5], finally leading to fully autonomous driving (AD), [6], will reduce harm to people, increase efficiency and reduce pollution.

Vehicles equipped with ADAS features started to be introduced in the 2000s. With these, a human driver is still fully responsible for the control of the vehicle. In the last few years, vehicles equipped with Automated Driving Systems (ADS) have started to be introduced. These have the capability to take over the responsibility for the control of the vehicle for parts of a journey. These are seen as a part of an inevitable development of the fully autonomous vehicle.

The Public Road Transport System

The Public Road Transport System (PRTS) is the public space used for automotive transport. This is a very rich environment with many different aspects. There are a variety of different types of users, who themselves are part of the PRTS. In addition to the road network itself, the PRTS has many different facilities that support and coordinate its users. The PRTS also includes broadcast information, the weather conditions and illumination levels.

The road network consists of many different types of road that traverse a variety of different settings from city to suburban to country. The road surface is made of different materials and also has words and symbols written on it. The road network has developed over centuries and while modern roads may be laid out in a geometric and somewhat logical manner, there still exits road dating back many centuries which are much less so.

Third Parties

In addition to the users of the road network there are many third parties involved in the wider system. Governments, and their agencies, create a legal framework for the design of the road network, the design and manufacture of the vehicles that use it and the behaviour of the people that are part of the PRTS. Standards bodies create standards for various other aspects of the PRTS including the design and construction of roads, vehicles and traffic information. The roads and vehicles are designed and built by different companies using many sub-suppliers and contractors. The roads and vehicles are maintained by organisations that did not design them. Broadcasters provide information used in planning journeys.

The Dynamic Driving Task and the Driving Enterprise

<u>The Dynamic Driving Task (DDT)</u>

Historically the task of driving a vehicle was not formally defined; it was a skill that had to be acquired. In most countries, there are measures to ensure that only those with a sufficient level of skill are allowed to drive and also measures to punish instances of poor driving. In the UK, the Highway Code was first launched in 1931 and, from 1934 onwards, passing a compulsory driving test was required before a driver was issued with a driving license. Legislation motivates the human driver to perform well by making poor driving an offence, examples include causing death by dangerous driving and driving without due care and attention.

More recently, vehicles have started to be equipped with ADS features which perform some aspects of the driving task. The term ADS itself is defined in SAE J3016, [7], as " *The hardware and software that are collectively capable of performing the entire DDT on a sustained basis, regardless of whether it is limited to a specific operational design domain*".

The term *operational design domain* (ODD) is defined by SAE J3016, [7], as: "*Operating conditions under which a given driving automation system or feature thereof is specifically designed to function, including, but not limited to, environmental, geographical, and time-of-day restrictions, and/or the requisite presence or absence of certain traffic or roadway characteristics.*" The ODD is effectively a subset of the aspects of the Public Road Transport System mentioned above.

With the advent of Automated Driving Systems (ADS) equipped vehicles, the driving task has started to be defined more formally

With the advent of ADS equipped vehicles, the driving task has started to be defined more formally. SAE J3016, [7], uses the term Dynamic Driving Task (DDT) to capture all the aspects associated with controlling the movement of a vehicle along a public road. These include:

- Determining the destination of the journey and the sequence of vehicle manoeuvres necessary to reach it
- Monitoring the environment external to the vehicle to detect objects and events and determining the correct vehicle manoeuvre in response to what is detected
- Implementing the vehicle manoeuvres by control of the lateral and longitudinal motion of the vehicle
- Ensuring other road users are aware of the presence, and manoeuvring intention, of the vehicle

The Driving Enterprise

The DDT definition captures the key aspects of controlling a vehicle from the perspective of designing a control system, however there is more to using a vehicle to make a journey than this. Other necessary aspects include:

- Ensuring that the vehicle is roadworthy before embarking on the journey
- Deciding on whether or not to embark on the journey, and the route to be taken, given the anticipated weather conditions
- Matching the demands of the external environment with the capabilities of the vehicle when driving
- Dealing with issues encountered during the journey regarding the vehicle, the road infrastructure and the weather
- Interacting with other human beings using, or in the vicinity of, the road

This wider view of planning and making a journey is referred to as the Driving Enterprise (DE) and its aspects are now briefly considered further.

Embarking on a Journey

The UK government recommends that prior to every journey the vehicle's tyre pressure, engine oil level, screen-wash level, lights and fuel level should be checked. Journeys should also not be started, or continued, if the demands on the vehicle would exceed the Vehicle Operational Constraints. Such constraints include ground clearance, wade depth, turning circle plus the minimum and maximum external temperatures under which it is designed to operate.

Given a roadworthy vehicle, there are other decisions related to the journey that need to be made. One is the decision to start the journey and the route to be taken, based on the purpose of the journey and the information available prior to starting. This may be advance information received as off-board data concerning road and weather conditions that may be encountered during the journey. The trustworthiness of this information needs to be evaluated given that it may be false, ambiguous, out of date or misleading.

Matching the Vehicle Capabilities with the External Environment Demands

To perform the DDT it is necessary for the environment external to the vehicle to be sensed and understood. At any one time, the ability to sense has limits. Trees, vegetation and buildings may obscure line of sight. Weather phenomenon such as heavy precipitation, fog and mist also affect visibility. It is important that manoeuvres are not attempted at speeds which are inappropriate for what can currently be sensed.

As mentioned, all vehicles have Vehicle Operational Constraints and some of these may be adversely affected when the vehicle develops a fault. When determining the correct vehicle manoeuvre, it is necessary to take into account the current performance capability of the vehicle so a manoeuvre is not attempted that the vehicle is not capable of making or will not complete in sufficient time. Examples including driving round a bend too fast and not having sufficient time to complete an overtaking manoeuvre.

It is also necessary to be aware of any reduction in tyre grip on the road, which may arise due to the poor quality of the road surface or the affect of the current weather conditions. This is significant because reduced grip on the road may make it difficult or impossible to complete a manoeuvre.

Managing Issues During the Journey

The DDT has to be performed in a road infrastructure that may not be "text book". The roads or junctions may be long standing and not designed to current standards. The road surface may be less than ideal, for example potholed or untarmacked with raised ironwork; as mentioned above, this may affect tyre grip. The placement of instructive and informative signs may make them difficult to read, they may also be obscured, graffitied or not lit. The content of information provided in signs may be out of date.

Once a journey has been started, there is a continuous process of decision making regarding whether to continue the journey on the same route, change the route, change the destination or stop the vehicle. Such decisions may be made in response to issues related to the weather, road infrastructure and vehicle faults.

A change of route or destination may be prompted by an awareness that the weather or road situation ahead could prevent completion of the journey by the current route or to the current destination. This awareness may be as a result of information received as off-board data or as a result of what is sensed directly. It may also be because of a vehicle issue, e.g. insufficient fuel.

A decision to stop the vehicle may be prompted by the onset of extreme weather, encountering significant road damage or the occurrence of a significant vehicle problem.

If the vehicle problem is such that it diminishes its roadworthiness, then a decision has to be made whether to stop the vehicle or to continue driving more cautiously with the intention of getting the vehicle checked at a garage at a later date. Whenever a decision to stop the vehicle is made, then it should be brought to rest in a location where no harm can occur. The vehicle also needs to stop if it is involved in an accident.

Legal and Social Aspects

The use of the vehicle on the public roads is a social activity that is governed by civil and criminal law. The laws prescribe particular behaviour and require a certain standard of performance. In the UK, the driver is expected to drive in the way that would be expected of a competent and careful driver such that the driving is never dangerous. To not do so may break the law regarding dangerous driving.

The act of driving also has socially expected norms of behaviour. It is generally expected that the vehicle behaviour will demonstrate a sense of care, courtesy and consideration for other road users and not act maliciously. In the UK, to not do so may break the law regarding careless or inconsiderate driving.

There is also the need to balance the obligation to abide by letter of the law with the need to infringe the law in those circumstances that require it in order to avoid harm, for example moving into a "bus only" lane to avoid a collision with a person unadvisedly crossing the road.

The directions given by authorised people, e.g. police officers, should be followed. These may be counter to the desire of the driver and/or may also infringe the letter of the law. Such directions will be being given to achieve a more important social good. A decision also needs to be made whether directions from an ordinary member of the public should also be followed. These may be to achieve a greater social good, but may also be intended for harm or have criminal intent. It is also necessary to allow passage for emergency vehicles, this may also require infringing the letter of the law.

DDT & the Driving Enterprise Performed by a Human Driver

The vehicle roadworthiness is usually the responsibility of the vehicle owner. The pre-journey checks should normally be carried out by the driver, but in the case of a hire vehicle, these are the responsibility of the hire company.

When planning and making a journey, the other aspects of the DDT and DE are the responsibility of the human driver of the vehicle. These include not attempting to use the vehicle if the demands on the vehicle would exceed the Vehicle Operational Constraints, and assessing the trustworthiness of any off-board data that is available.

During a journey, all the sensing, understanding and decision making related to vehicle movement is the sole responsibility of the human driver. They are also responsible for handling any mechanical or electronic control problems with the movement control systems.

The expectation that the human driver can match the vehicle capabilities with the external environment demands is typically based on the fact that the driver is required to be trained. This may be coupled with being licensed to drive, and the fact that the driver is required to be fit to drive, including having adequate eyesight.

Human beings are able to perform the driving task, cope with problems and interact socially with other human users of the road because they possess general intelligence and a theory of mind. Mitchell, in her 2020 article, [8], notes that human understanding is built on a foundation of innate core knowledge of physics, gravity, object persistence, and causality. This enables humans to trace the relations between objects and their parts, think about counterfactuals and what-if scenarios, and act in the world with consistency. They are able to apply previous experience and learning to new situations. Possessing a theory of mind means that human beings have empathy for other human beings and gives them insight into their future behaviour and actions.

DDT & the Driving Enterprise Performed by a Level 5 Vehicle

SAE scale defines a 6 level taxonomy of driving automation [7]. On this scale, a fully autonomous vehicle is referred to as a Level 5 Vehicle. In such a vehicle, the ADS is responsible for the sustained and unconditional performance of the entire DDT, including when the ADS itself experiences faults or failures. Under no circumstances is the user expected to respond to a request to intervene, and there is no remote operator who can intervene.

As there is no action required of the user, the Level 5 Vehicle has to, not only perform the DDT, but also, address the full Driving Enterprise, while taking into account the Vehicle Operational Constraints. This includes making all the decisions related to the journey, as noted above.

The case where the vehicle is sharing the road with other users, including non-autonomous vehicles, 2 wheeled vehicles, pedestrians and animals, both wild and domesticated, is now considered.

Planning and Continuing a Journey

The pre-journey checks still have to be made by the owner/user of the vehicle prior to commencing the journey.

For a privately run vehicle, the owner, and/or user, decides the destination and any waypoints. If the Level 5 Vehicle plans the route, then it has the responsibility of taking account of off-board data, e.g. current traffic and weather conditions, and assessing its trustworthiness. It should not attempt to make demands on the vehicle that would exceed the Vehicle Operational Constraints.

The decision to start a journey, and continue a journey, in the light of off-board data and the situations encountered during the journey, is the responsibility of the Level 5 Vehicle. An incorrect decision may put the occupants lives at risk.

Vehicle maintenance

The vehicle owner still retains responsibility for ensuring the roadworthiness of the vehicle. Although, not having to perform the driving task themselves, may diminish their sense of responsibility as they will not directly experience any difficulties in controlling the vehicle.

If the ability of a human driver to detect mechanical problems, via mechanical vibration, a change in driving feel or a change in vehicle performance, is to be retained, then additional sensors and algorithms may be required. An alternative may be enforcing a more rigorous maintenance regime.

The underlying mechanical components currently in use have been developed for reliability and length of operational lifetime based on a human driving style profile. If the style of driving performed by a Level 5 Vehicle is significantly different to that of a human driver in particular aspects, e.g. greater use of friction brakes, then this may result in faster wear out of associated mechanical components. If this is the case, then the design of these components may need to be changed. Attempting to mimic the driving style of a human driver may be seen as too difficult or too restrictive.

Executing a Journey

All the responsibilities that would traditionally be with a human driver, are now the responsibility of the Level 5 Vehicle. It must discharge these without any human intervention in such a way that no human is injured; this includes both this vehicle occupants and all other road users.

As mentioned previously, these responsibilities are sensing the environment external to the vehicle and executing the appropriate manoeuvres while taking account of the Vehicle Operational Constraints and the demands of the road. Also, interacting socially with other human road users and handling problems presented by the weather, the road and vehicle faults.

The sensing capability of a Level 5 Vehicle may exceed that of a human driver in terms of having multiple sensors that may have a greater range and resolution and may not be degraded by low levels of illumination or by the presence of moisture in the air. However, understanding what is sensed does not have the benefit of general intelligence. A Level 5 Vehicle may also benefit from V2X (vehicle to everything) communications which provide information about the immediate environment external to the vehicle, including the disposition of other vehicles, [9], [10]. The trustworthiness of this information would need to be assured, as it is outside the remit of the Level 5 Vehicle.

Such "artificial intelligence" does not embody a theory of mind

A human driver utilizes their general intelligence to execute the journey, as mentioned above, while the development of Level 5 Vehicles draws heavily on machine learning. Such "artificial intelligence" does not embody a theory of mind; it is not based on an innate core knowledge of physics, gravity, object persistence and causality. Rather, it is based on a correlation between a huge number of data points. It does not form abstractions, which means that it cannot extrapolate previous experience onto novel situations. However, the legal threshold for acceptable driving performance is still currently based on what can reasonably be expected of a competent and careful human driver.

With a human driver, it is only the movement control systems that can experience technological failures, however, with an autonomous vehicle, the systems responsible for the autonomy can also experience technological failures. The software responsible for this may be hosted in a movement control system. Without the recourse to a human being, the vehicle has to handle both types of failure without harming anyone.

One can understand why the use of driverless vehicles is currently in closed environments and low speeds

From these considerations, it can be seen that producing a Level 5 Vehicle to operate in an unrestricted environment is going to be a significant challenge. It is not surprising that the current general road systems are only at Level 2 or Level 3, despite the predictions of a few years ago, [11]. One can understand why the use of driverless vehicles is currently in closed environments and low speeds as this removes many of the difficult issues. This is likely to remain the case for somewhile.

Level 5 Vehicle Deployment in an Unrestricted Environment

The implications of having a Level 5 Vehicle deployed in the current unrestricted environment, and the challenges it represents, have been described. If a Level 5 Vehicle is to be deployed in this way, and also have a credible safety argument, it may be better to take a different approach from just replacing a Level 0 Vehicle with a Level 5 Vehicle while leaving everything else the same.

Two different approaches that could be taken are briefly discussed. These are changing the infrastructure to be more accommodating to the ADS and extending the scope of the safety argument.

Infrastructure Change

Rather than trying to develop ADS vehicles that can work in the current road environment with its mixture of modern roads developed for human drivers, and historic road infrastructure, should it start to be developed so as to take into account the strengths and limitations of the ADS? Some possibilities are briefly considered.

The standards governing the construction of the road infrastructure, for example, junction design and signage on the road surface and on free standing signs, is currently based on the capabilities of humans. There may be opportunity for the standards to be optimised for the sensing capabilities of an autonomous vehicle. This may affect the placement of signage and the rendering of the information it conveys. It may also affect the layout of junctions and their associated road markings. If the safe operation of autonomous vehicles depends on the road infrastructure being constructed and maintained in accordance with specific standards, then there is greater obligation on those third parties responsible to adhere to those standards.

There may be an opportunity to optimise the rules of the road such that it is easier for autonomous vehicles to comply with them, and also to anticipate the behaviour of other autonomous vehicles. Such changes would be the responsibility of the government. The government may also have a role to play in licencing the operation of each model of autonomous vehicle in the same way that it currently licences human drivers.

Changes such as these could then be taken as the basis for assumptions in an extended safety argument as discussed in the next section.

Extending the Safety Argument

Currently the safety arguments that are produced for ADAS and ADS features for vehicles with human drivers, whether by the OEM or by a tier 1 supplier (company that supplies major parts or systems directly to an OEM), are purely a technical argument about the particular vehicle feature. The other aspects of the road transport system, (see section above entitled Third Parties), are deemed not to be in the scope of the argument, which is purely for the hardware and software installed on the vehicle. This is the scope required by ISO 26262 – Road vehicles – Functional Safety, [12]. Any assumptions concerning other aspects of the road transport system are usually implicit and not documented.

There is also an undocumented assumption that the human driver can cope with, and resolve, any issues arising from the actions, errors, etc, associated with all of the third parties. This is supported in law by the criteria – "was the behaviour of the driver what can reasonably be expected" as assessed by the driver's peers. That this is not an unreasonable expectation of a human driver was discussed, (see section above entitled DDT & the Driving Enterprise performed by a human driver), as were the difficulties that it presents to a Level 5 Vehicle.

An alternative approach could be for the safety arguments produced for ADAS and ADS features to explicitly identify the assumptions made concerning all relevant third parties. This would provide a greater understanding of the dependencies between third parties. It would also prompt the documenting of assumptions that one party makes about the responsibilities of others. For this to be beneficial, these third parties would need to state explicitly what can be assumed by the safety arguments produced for ADAS and ADS features and where appropriate, for those third parties to provide their own argument supporting those assumptions.

Based on what was discussed in the previous section about the infrastructure, topics about which assumptions might be made include:

- The rendering and placements of signs
- The physical limits of the roads, e.g. widths, radii, and the road markings
- Those aspects covered by the government licensing of vehicles
- The behaviour of other road users, based on the rules of the road

Some aspects of these may be stated in the existing standards, but currently they are not explicitly referenced as assumptions in the safety argument. It is also the case that some aspects vary from country to country.

It is the vehicle manufacturer who is ultimately responsible for the safety argument for the ADAS or ADS feature, as integrated into the vehicle. At present, organisations in the supply chain, particularly the tier 1 suppliers, do provide information to the vehicle manufacturer. This is on an *ad hoc* basis, and the content of the information is often subject to intellectual property rights considerations. The provision of information necessary for a strong safety argument could be made an enforceable obligation.

This enforceable obligation could be extended to the maintainers of roads and vehicles and also to the providers of information updated in real-time. Topics about which assumptions might be made include:

- The reliance on road maintainers to maintain the roads to the construction standard, provide adequate communication of a problem or close the road if necessary
- The reliance on maintainers' ability to leave the vehicle up to the manufacturer's specification after having performed a maintenance task
- The reliance on the providers of information updated in real-time for the information to be available, up to date, accurate and clear. Such information may concern possible and actual weather conditions or state of the roads.

Some of the changes discussed here may also support driving automation levels 3 & 4. These have the added complication, not addressed here, of the driver participating in the hand-over of the DDT with the vehicle, both to and from. They also have the complication of the human driver being responsible for monitoring the performance of the ADS or ADAS feature and intervening when necessary.

Concluding Remarks

The article has described how planning and completing a journey on the public roads in a vehicle, referred to as the Driving Enterprise, is more than just performing the Dynamic Driving Task, but also, involves many other activities. While the consequences for autonomous vehicles of these other activities are being mentioned by some other researchers, the main focus seems to be on developing particular technologies for the vehicle. It is hoped that this article has brought these issues to the forefront of more people's minds.

As a small contribution, two key ideas have been presented that could help progress the field toward full automation. The first is to change the design of roads and signage to better suit autonomous vehicles, without disadvantaging non-autonomous vehicles. The second is to include, in the safety argument for an autonomous vehicle, more of those aspects of the external world on which the vehicle, implicitly or explicitly, relies.

If there is an overall conclusion, then it is just that it might be best not to deploy a Level 5 Vehicle in the current environment. The roadmap from where we are now to a Level 5 Vehicle deployed in an unrestricted environment with all other road users is likely to be long and tortuous and beyond the scope of this article.

References

[1] University of York. (31/01/2022). Assuring Autonomy International Programme. Available: https://www.york.ac.uk/assuring-autonomy. Accessed April 2022.

[2] R. S. Rivett, "Public Road Transport System and Vehicle Models," University of York 2022. https://www.york.ac.uk/assuring-autonomy/research/publications/. Accessed April 2022.

[3] MISRA. (29/01/22). MISRA Safety Arguments. Available: https://www.misra.org.uk/misra-safety-argument. Accessed April 2022/

[4] A. Ziebinski, R. Cupek, D. Grzechca, and L. Chruszczyk, "Review of Advanced Driver Assistance Systems (ADAS)," presented at the AIP Conference, 2007.

[5] "Advanced driver assistance systems," European Road Service Observatory2018.

[6] A. Faisal, T. Yigitcanlar, M. Kamruzzaman, and G. Currie, "Understanding autonomous vehicles: A systematic literature review on capability, impact, planning and policy," Journal of Transport and Land Use, vol. 12, pp. 45-72, 2019.

[7] SAE, "J3016 Taxonomy and Definitions for Terms Related to Driving Automation Systems for On-Road Motor Vehicles," ed: SAE, 2021.

[8] M. Mitchell. (2020) On Crashing the Barrier of Meaning in AI. AI Magazine. 86–92.

[9] L. Hobert, A. Festag, I. Llatser, L. Altomare, F. Visintainer, and A. Kovacs. (2015) Enhancements of V2X Communication in Support of Cooperative Autonomous Driving. IEEE Communications Magazine.

[10] C. Jung, D. Lee, S. Lee, and D. H. Shim, "V2X-Communication-Aided Autonomous Driving: System Design and Experimental Validation," MDPI Sensors, 2020.

[11] SingularityHub. (2013, 29/01/2022). Toyota Joins Slew of Major Automakers Promising Self-Driving Technology This Decade. Available: https://singularityhub.com/2013/10/31/toyota-joins-slew-of-major-automakers-promising-self-driving-technology-this-decade/. Accessed April 2022.

[12] ISO, "ISO 26262 Road vehicles - Functional safety," ed, 2018.

Roger Rivett, Independent

Roger spent over 25 years working on functional safety for an automotive manufacturer. He was a founder member of MISRA and was its chair for 15 years. He was a member of ISO-TC22-SC32-WG8 from 2005 until 2018. He is a Program Fellow on the University of York Assuring Autonomy International Programme and a member of the MISRA Automotive Safety Argument working group.

THE SAFETY-CRITICAL SYSTEMS CLUB

Seminar: Managing Unexpected Risks

Handling Rare and Severe Events (Black Swans) Now and in the Future

26th May 2022, BCS London and blended online

This seminar will consider the nature of major and unexpected risks and how to plan for, assess and manage recovery in a safety context. Black Swans are events which are rare, 'out-of-blue', have high impact but are explainable after they have happened. Major examples might be the terrorist attacks of 9/11 or the loss of Malaysia Airlines 370.

There are many aspects to the management of such events including planning, preparedness and dry-runs of contingency processes. When an event occurs, it is necessary to quickly establish the nature and scale of the problem, stabilise the situation, prevent of a cascade of failures, assess risks, provide a contingency service if possible, communicate to all stakeholders and eventually recover normal operations.

Communication, obtaining reliable status information and rapid assessment of risks are critical but may be difficult. Hard data may be limited, and situational awareness, human factors, organisational experience and safety culture all come into play.

This seminar looks at the position in various industries. There will be a discussion session on how the situation changes when autonomous functionality is involved. How do we make rapid judgements when human involvement is small?

There will be a Q&A session where delegates can explore the concept of Black Swan events and possible approaches to dealing with them.

This event will be held at the BCS (The Chartered Institute for IT) at 25 Copthall Ave, London EC2R 7BP.

How to Manage Unexpected and Severe Events

Bookings at: www.scsc.uk/events

This seminar is an opportunity to hear about management of rare and high impact events across different industry sectors and how this is likely to change in the future.

It will be useful for safety practitioners, safety managers, and for those involved in the planning and management of high-impact events.

Details at: www.scsc.uk

www.scsc.uk

Brexit and Software Safety

Dai Davis, is a Technology Lawyer with his own specialist law practice, Percy Crow Davis & Co. For over three decades, Dai has been practising in high-tech product safety, including advising on safety related software, CE Marking and product recall. In this article, Dai discusses the impact Brexit has had on three important areas of law dealing with software safety.

I recall reading years ago about what the writer said was the first recorded death caused by software, or perhaps I should say, badly programmed software. It was the story of an unfortunate worker in a car factory in New York State in the early 1960's being welded to death by an automatic robot welding machine. This was not a pleasant death, I imagine. Incidentally, if anyone has more details of this or another "first recorded death by software" story, I would be pleased to hear it.

It has taken a long time to see snippets of law, and they are only snippets of law, deal specifically with software safety, but interestingly those that are found in law are changing after Brexit. In this article I will discuss three areas of divergence in a post Brexit world:

- CE Marking which governs product safety
- the freedom of the United Kingdom to make new un-harmonised law, the example I will use is of car charging port legislation
- database rights.

For each of those areas I will explain the changes and the link to safety related software.

European Union Product Safety Legislation

Software is found in a multitude of products. In the European Union, as part of what was then called the "1992 process", freedom of movement of goods was introduced. In practice it was later in the 1990's before much of the necessary European legislation came into force. The primary legislation underpinning this process is the "CE Marking" legislation, which requires most, though not all, products to bear the "CE Marking" and adhere to certain qualitative criteria.

The United Kingdom has taken many steps to ensure that appropriate changes were made to legislation in the United Kingdom to smooth the transition to Brexit. For political reasons, the European Union did not take the equivalent steps. Therefore, in almost all areas affecting commerce, we have what is commonly termed a "hard Brexit".

The area covered by the CE Marking Directives is a good example of the effect of this hard Brexit. The United Kingdom amended the CE Marking legislation by the "Product Safety and Metrology etc. (Amendment etc.) (EU Exit) Regulations 2019" regulations [1], which run to some 793 pages. The European Union made no changes to its equivalent legislation. As a result, this means, for example, that we now have two[1] entirely separate CE Marking regimes, one for the United Kingdom[2], the other for the European Union.

One of the requirements of the CE Marking Directives is that a manufacturer must meet the appropriate standards. There is a hierarchy of standards:

- if there is a European standard formally adopted by the European Union, that standard must be followed
- otherwise the relevant international standard (if there is one)
- otherwise the relevant national standard.

However the hierarchy now to be followed in the United Kingdom is:

- if there is a British standard formally adopted by the United Kingdom government, that standard must be followed
- otherwise the relevant international standard (if there is one)
- otherwise the relevant (not formally adopted) British national standard.

At the moment there are no significant differences in these hierarchies of standards. Furthermore, the British have always pulled far beyond their national weight when it comes to influencing both European and International standards. The United Kingdom has also been allowed to retain its seats on the European standards making bodies: CEN and CENELEC. While eventually there will be divergences between the United Kingdom and European Union standards regimes which underpin the CE and UKCA Marking legislation, those differences will take a long time to come to the fore.

[1] Actually, this is incorrect, as we will have three regimes: one for Great Britain, another for Northern Ireland and a third for the European Union. However, a discussion of the special regime for Northern Ireland and the associated UKNI Marking is outside the scope of this article.

[2] In the United Kingdom the product will, if it is placed on the market on or after 1 January 2023, need to bear the UKCA Marking rather than the CE Marking.

Importantly, the Electromagnetic Compatibility legislation requires that electronic equipment meets qualitative criteria to ensure that it does not unduly emit electromagnetic interference and that it is immune from undue influences from external electromagnetic sources. Since software must reside on an electronic device, that means that the device must comply with that legislation. Having said that, there are no explicit provisions in the Electromagnetic Compatibility legislation dealing with software. However, that is not the same in respect of all the CE Marking legislation. In particular, the Machine Safety legislation and the Medical Device legislation both contain explicit provisions regarding software. By way of example I look at the former in detail below.

Machine Safety legislation

There are two basic requirements which must be satisfied in order to comply with the Machine Safety legislation. First, the legislation contains a list of specific attributes that must be satisfied, second, the relevant machine must satisfy the appropriate standards.

Annex 1 of the machine safety legislation contains a list of those specific attributes[3] and clause 1.2 of the Annex contains the following reference to software:

"Safety and reliability of control systems: Control systems must be designed and constructed in such a way as to prevent hazardous situations from arising. Above all, they must be designed and constructed in such a way that ...

- *a fault in the hardware or the software of the control system does not lead to hazardous situations, [and]*
- *errors in the control system logic do not lead to hazardous situations, ...".*

In addition, where machinery includes a "logic unit to ensure safety functions", a higher level of certification procedure is sometimes required by the legislation.

Ultimately the United Kingdom and European Union interpretation of this identical legislation will change. That will also affect software safety. For example, there have been previous unsuccessful attempts in the United Kingdom to require that safety-related software is written only in approved software languages. If there are such attempts in the future, it would be much easier for them to succeed since they would not need also to be adopted by the remainder of the European Union. The formation of the new UK Cyber Security Council is a good example of the direction in which the United Kingdom is travelling.

Legislation specific to the United Kingdom

Previously the United Kingdom has been unable freely to legislate in many areas of law. Now, after Brexit, it has much greater freedom to do so. An example of United Kingdom specific legislation appearing, which would not have been as easily adopted if the United Kingdom had remained within the European Union, is the Electric Vehicles (Smart Charge Points) Regulations 2021 [2].

[3] Referred to in the legislation as "Essential Health and Safety Requirements".

Schedule 1 to those Regulations, which must be complied with where a charging point is sold on or after 31 December 2022, imposes (partly software) security requirements for those charging points.

While a full discussion of the relationship between software safety and software security is outside the scope of this article, suffice to say that one cannot have one without the other. By way of example, the charging point must be configured so that the software can be securely updated[4]. One danger if this does not occur could be that the charging point, particularly since some charging points come with large external electricity storage devices, becomes incompatible with the local electricity grid and damages it. While human injury might be relatively fanciful in this example, software safety is not, in law, limited only to injury to humans. Safety in law includes safety to goods, and electricity is goods for these purposes[5].

Database Rights

A less direct interaction between Brexit and software safety is in the field of databases. These days the boundary between what is software and data has become increasingly blurred. At one level, one can regard any safety related software as merely data, since it is data which is input into a compiler to produce executable code[6]. An obvious example of a safety related database would be a database in a hospital storing patients' blood types or allergies. Correspondingly, a software program to store, retrieve and manipulate data in that database would be safety related.

Databases are protected by a European Union database right. Unlike most other intellectual property rights, which are based on international conventions, database rights are a peculiarly European invention. Under this legislation, the maker of a database (i.e., a person who creates a database) has the right to prevent the extraction or re-utilisation of the whole or a substantial part of the contents of the database. The legislation came into force on 1st January 1998. As indicated, this legislation is specific to the European Union and is not based on an International Treaty.

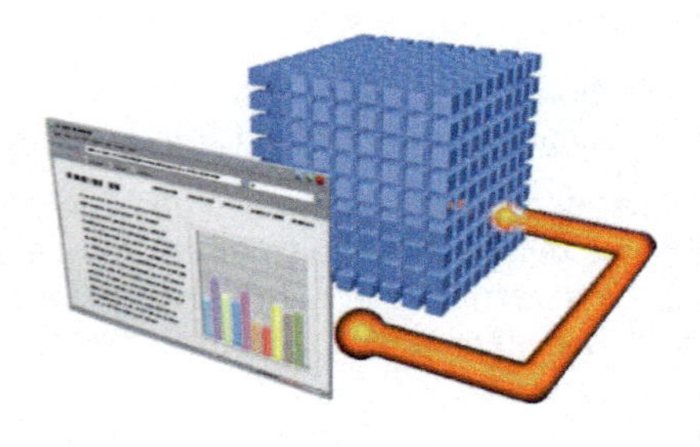

For these purposes, a database must have an element of selection in it as well as an investment of time and money to create it. For example, a database could consist of the events in a football match, such as the free kicks, corners, penalties and goals, or a database could consist of a list of medicines together with an overview of what the attributes of and indications for each medicine are.

The rights created by the European Union legislation last for 15 years from the date of creation of the database.

[4] The Regulation provides a degree of detail as to how this may and may not be achieved.

[5] See, for example, the express provision in Article 2 of the 1985 Product Liability Directive [3].

[6] Although such a data set (i.e. a computer program) would not satisfy the technical, legal definition of what constitutes a database for the purposes of the database legislation.

The rights are infringed even by a systematic extraction or re-use of insubstantial parts of the contents of the database. Where there is a substantial change to the contents of a database, so that it can be considered as a "substantial new investment", the database will then qualify for a new term of protection beyond the original 15-year period. In this way, database protection can conceivably last for a long time, provided that "substantial new investments" are regularly made to the database. This will invariably be the case for commercial databases which are continuously being updated.

The key is that there must have been a substantial investment in the obtaining, verification, or presentation of the contents of the database. A further condition is that, in order for the database right to be enforceable in the European Union, the person or organisation who is the creator of the database must be a national of a Member State of the European Union or a company formed under the laws of a Member State and based there from an economic perspective.

The important thing to notice in this condition is the use of the present tense in the phrase commencing "be a national of a Member State".

Where the person or organisation which owns the database, rights, ceases to be a national or company of a Member State, the database right is lost. So, if a company relocates itself and assets to a country which is not a Member state, such as the United States, the database rights are lost.

Unfortunately, it is not absolutely clear that the ambiguous Brexit deal that has been negotiated ensures that United Kingdom citizens and companies will retain their database rights now that the United Kingdom no longer is a Member State. While the intention of the Brexit deal was clearly that United Kingdom citizens and companies should retain their Database Rights, there is at least one counterargument to that position. That stems from the simple fact that the European Union legislature has not, in fact, changed the wording of the Directive.

Conversely, within the United Kingdom, European Union citizens and companies, alongside United Kingdom citizens and companies, will continue to have enforceable Database Rights. This continuation is specifically and unambiguously provided for in the initial Withdrawal Agreement which was agreed in January 2020.

The position of United Kingdom owned databases is therefore at best unsatisfactorily ambiguous, and at worst, there is no continuing protection for United Kingdom databases in Europe.

However, for United Kingdom companies and citizens, the reverse position is more complex, and the position cannot be stated with absolute certainty. It is the associated doubt itself which has created a disadvantage for United Kingdom companies. In a situation where a United Kingdom citizen or company seeks to enforce its database rights in Europe it may find itself coming up against an argument that, in fact, it no longer has database rights in Europe.

Not only this, but the enforceability may also depend partly upon the country instantiation of the Member State in which that individual or company is seeking to enforce its database right. The European legislation is found in a 1996 Directive.

All Directives require a country instantiation in order to be enacted in that country. In some countries, the wording of the Directive states that it is sufficient only that the database owner resided in the European Union at the time the database was created. For example, this appears to be the case in the German instantiation of the Directive.

The position of United Kingdom owned databases is therefore at best unsatisfactorily ambiguous, and at worst, there is no continuing protection for United Kingdom databases in Europe.

Either way, it does not reflect the agreed aim in the political declaration entered into between the European Union and the United Kingdom in October 2019. That declaration included an obligation "to preserve the Parties' current high levels of ... rights [in] ... database[s]".

Furthermore, database rights are becoming more, not less, important. This is because of the use of artificial intelligence to create those databases. By way of an example, consider a computer generated database of protein molecules and their likely properties. It is perhaps unfortunate, to say the least, that United Kingdom companies appear to be at a legislative disadvantage in the future exploitation of databases.

References

[1] The Product Safety and Metrology etc. (Amendment etc.) (EU Exit) Regulations 2019, https://www.legislation.gov.uk/uksi/2019/696/contents, accessed April 2022.

[2] The Electric Vehicles (Smart Charge Points) Regulations 2021, https://www.legislation.gov.uk/uksi/2021/1467/contents/made, accessed April 2022.

[3] Council Directive 85/374/EEC of 25 July 1985 on the approximation of the laws, regulations and administrative provisions of the Member States concerning liability for defective products, https://www.legislation.gov.uk/eudr/1985/374#, accessed April 2022.

Dai Davis

Dai Davis is a Technology Lawyer. He practices as a solicitor but is also a qualified Chartered Engineer and Member of the Institution of Engineering and Technology. Dai has consistently been recommended in the Legal 500 and in Chambers Guides to the Legal Profession for over 25 years. He has two master degrees: one in Physics, the other in Computer Science.

Having been national head of Intellectual Property Law and later national head of Information Technology law at Eversheds for a number of years, Dai has for the past decade been a partner in his own specialist law practice, Percy Crow Davis & Co. Dai advises clients throughout the country on intellectual property, computer and technology law subjects. A third "super specialism" that Dai has practised for over three decades is high-tech product safety, including advising on safety related software, CE Marking and product recall.

He is primarily a non-contentious lawyer, advising clients on technology-related commercial agreements. Being a technologist, Dai often works with clients who are involved in cutting edge technologies.

Dai is a non-executive director of FAST (The Federation Against Software Theft) and a Liveryman of the City of London through the WCIT (Worshipful Company of Information Technologists). He has been a Council member of the Licensing Executives Society of the United Kingdom, a body of professionals dealing in intellectual property licensing, for more than a decade. He is widely published and an experienced public speaker and writer. He can be contacted on 07785 771 721 or mail@daidavis.com.

SSS'22 Event Report

The symposium had a special flavour this year, celebrating the club's successes over the years and giving a taste of what the next 30 years in safety engineering might be like. After being forced to run entirely online last year, the Safety-Critical Systems Club's annual Symposium finally returned to an in-person event hosted at the Bristol Royal Marriott in February. This was a blended event with significant online attendance, and a mix of physical and virtual presentations – the first symposium of its kind in the club's 30-year history.

The outbreak of the Covid-19 Omicron variant towards the end of 2021 certainly put the event in jeopardy, and it was only in January 2022 that a firm decision was made to go ahead with the event. Despite the uncertainties, the event was remarkably well-attended, with approximately 120 delegates being split equally between in-person and online attendees.

Neither did circumstances detract from the quality of the presentations with excellent talks being covered in areas such as new techniques and applications, human factors, assurance, autonomy, healthcare and safety & security integration.

There was also a bumper pack of merchandise and publications for those attending in person – as well as the symposium proceedings, there were several working group guidance publications, the Feb 2022 edition of the SCSC Newsletter, the 2021 annual compendium of newsletters, and the 30 Years of Safety Systems, which included a selection of important articles published in the Newsletter over the last 30 years, with additional commentary and postscripts from some of the original authors.

Shaping the Evolution of Safety Engineering

Professor John McDermid, Director of Assuring Autonomy International Programme at the University of York opened the conference with a talk on the future evolution of safety engineering. He introduced three concepts that have a bearing on system safety: Consumerism, Contradictions and Counterfactuals:

- Consumerism: eg. public buying power can influence the products in the marketplace, but these might not be the safest or most environmentally friendly
- Contradictions: eg. cars are considered safe and widespread, but contribute to other "Excess Deaths" through air pollution
- Counterfactuals: eg. Assessing the potentially incommensurable benefits and harms that would arise if an event had, or had not, occurred.

John said that safety engineering needs to look at taking a wider stance, to consider not just the technical aspects of safety, but also to incorporate the benefits and harms to individual, society and the environment. The growing complexity and autonomy of AI and ML poses huge challenges to safety engineering and there is a need to develop new techniques and extend existing ones. John provided some examples, such as extending safety arguments to include an ethical argument, methods for measuring and comparing complex systems and the use of digital twins in exploring counterfactuals.

John concluded that safety engineering is at a crossroads, both technically and politically. Technically from the need to broaden the discipline and consider ethics and sustainability and politically, seeking to raise awareness and influence policy through the mantra of "Safe, Ethnical and Sustainable".

Stories and narratives in safety engineering

Catherine Menon, Principal Lecturer at the University of Hertfordshire, illustrated the role of storytelling in constructing the narrative for a safety case. Through a brief retelling of the Hansel and Gretel tale, she illustrated how a story can have many different meanings and how different interpretations can be drawn.

Catherine gave Use Cases as a common form of narrative and described these as "fluffy white clouds", where everything works as planned and as expected. However, in real life, things go wrong and a narrative could also be in the form of "war stories" or "black clouds". Catherine asked what narrative form a Safety argument should take, and argued that a claim that a system is safe is actually a "white cloud" narrative, for example, it may not place a focus on counter-evidence and lack of traceability or rigour.

Other forms, such as justifying the author's decision that the system is safe, tend more to the opposite "black cloud" style of argumentation. Catherine said that a safety argument should be more about exploration, open to the possibility that it may or may not be safe, with the reader and writer working together to explain, not justify.

The readership of the safety case is also diverse, for example, it could be other engineers, regulators, users, or the general public and so needs to be framed in a way that they will understand. In particular, the way we communicate to the public can encourage them to think in particular ways eg. The aircraft landing in the Hudson is presented as a triumph in contrast to the Nimrod accident report that apportioned individual and collective responsibility.

Catherine warned about the pitfalls of storytelling and how powerful stories can be, as people can believe emotions over facts. She illustrated this through references to a legal case and other safety-related accidents where, how the story was told, led to a disproportionate allocation of blame to those involved.

She concluded by saying that we should still tell stories, but we should tell those stories safely.

30 Years of Safety Quiz

The first day concluded with a special 30th Anniversary Quiz hosted by Catherine Menon and Paul Hampton. The quiz comprised 19 questions on all matters to do with system safety and the SCSC, and was open to both in-person and online attendees working individually or in teams.

Questions were wide ranging from anagram solving, picture questions, multiple choice and concluded with a challenge for contestants to create a Limerick or Haiku.

Examples of some of the contributions are as follows:

Limericks

There once was a safety engineer,
Who lived life constantly in fear.
His friends all laughed,
And said he was daft,
"Stop stressing and drink lots of beer!"

There once was a girl called Stacy,
Who had a complete disregard for safety.
She chased all sorts of thrills,
With things that could kill,
And no one has seen her lately.

Haiku

Safety fans beware
Incidents are everywhere
Common sense is rare

Blossoms are falling
So is the plane I am on
Safety case was wrong

The winners of the quiz were Dewi Daniels and Roger Rivett with 66 out of 104.

You can try the quiz by following this link: https://scsc.uk/re797.24.

Answers can be found here: https://scsc.uk/re797.25

Events

As well as a number of ad-hoc events organised by delegates themselves through the Whova app, an organised tour of Bristol was arranged for delegates on Wednesday morning. This was a walking tour and took in some of the sights of Bristol such as the Bristol old Vic and Bristol Cathedral.

Ethics and Safety for Connected and Automated Vehicles

Paula Palade, from Jaguar Land Rover, discussed the status of various reference frameworks that are being developed to cover the ethics of Connected and Autonomous Vehicles (CAVs). One such framework was funded by the European Commission and Paula, as a member of an independent expert group developed 20 recommendations on road safety, privacy, fairness, explainability and responsibility. The purpose of the report was to explore some important questions:

- How safe should CAVs be?
- Are pedestrians and cyclists more at risk with CAVs in traffic?
- Do you need to understand the technology behind it?
- What kind of data will a CAV share?
- Can the decisions of a CAV be trusted?
- Who is responsible for its behaviour?

The report provided 20 new recommendations for a safe and ethical transition towards driverless mobility. Each recommendation was accompanied by more specific guidance, and Paula illustrated this by discussing some of the recommendations in more detail – for example:

- **Recommendation 1:** Ensure that CAVs reduce physical harm to people. The report recommended that data and statistics around collisions and near misses should be widely shared as there is usually lack of details about the exact circumstances of a collision and events are rare;
- **Recommendation 2:** Prevent unsafe use by inherently safe design. Manufacturers, deployers, together with researchers, should create intuitive, user-centred systems that are designed to prevent unsafe use.
- **Recommendation 3:** Define clear standards for responsible open road testing. A comprehensive and rigorous framework for open road testing would be most appropriately addressed at European level.
- **Recommendation 4:** When to break the rules... In some cases it may be safer to allow the CAV to break the road rules eg. to mount the pavement to allow an emergency vehicle to pass. Paula referred to promising work in this area using "ethical goal functions" as part of a hybrid AI system.

Paula also touched on other security related aspects, such as, issues around data privacy where many vehicle systems will store personal data that may be accessible by 3rd parties, and systems that may use a driver's behavioural and movement patterns for commercial gain.

Paula concluded by referencing IEEE Std 7000-2021: *Standard Model Process for Addressing Ethical Concerns during System Design,* and said this gives a high-level view of how to include ethical considerations in the design process, but lacks the linkage to how to translate ethics into actual non-functional requirements. She said that *BS ISO 39003 Road Traffic Safety (RTS) – Guidance on safety ethical considerations for autonomous vehicles* is currently in development and is intending to provide more detail in this area.

The report itself can be download from: https://europa.eu/!VV67my

The German and Belgian Floods

Peter Ladkin of Causalis, concluded the second day with a virtual presentation on the exceptional flooding that occurred in July 2021 that claimed the lives of over 220 German and Belgian citizens and caused over 33 billion Euros of damage.

Peter said the primary issue was that the "problem" rivers had large catchments in mountainous areas and flowed down into populated high-sided valleys. There were also problems even on flat land with human activities worsening the effects with "sealed soil" (tarmacked and built-up areas) and also in one case, a gravel pit excavation was suspected of leading to subsidence and landslide.

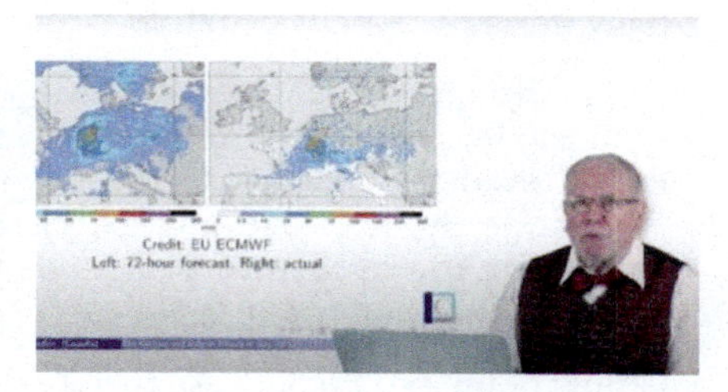

The floods were predicted 3 days in advance but the location of rainfall is not generally predictable to within 100km, and the predicted rainfall in some locations was as wide as 20mm-150mm.

Peter highlighted that one of the problems was on communication as some people were not aware of the severity of the floods in time to be effectively evacuated. He noted that the communication and disaster response is set along a political organisational hierarchy and not one better suited to disaster alerting and response. Peter suggested a different organisational model that might improve the management of such disasters and suggested that this model (which would require a change in German law) may have led to a quicker response and hence, better outcomes from the flooding that took place.

Peter concluded with some suggestions on improving outcomes from such events such as:

- education and training for residents
- local coordination and organisation of physical-aid
- rethinking approaches to "sealed soil" and ultimately
- arresting climate change to reduce occurrences of these types of extreme weather events.

Introducing a Restorative Just Culture and the Learning Review at the Docklands Light Railway

Adam Johns is a Senior Health & Safety Coach at Keolis Amey Docklands, franchise operator of the London Docklands Light Railway (DLR).

Adam said that when it comes to safety on the front line, many organisations are poor at learning from incidents as they do not treat the people involved in the right way to get the right sort of information from them to promote learning.

Adam provided an overview of the DLR – an automated light railway that has multiple different lines but is not fully automated – it has an attendant on board to operate the doors and deal with passenger issues.

Adam said that in 2019 as a result of a number of safety culture surveys, the organisation thought its safety culture was static and not progressing. A new Safety Director embarked on a programme called "Next Platform" to implement a revised safety approach based on Sidney Dekker's "Safety Differently" concepts. Next Platform's initial three phase are:

- **To change the conversation**: adopting a different non-pejorative and neutral language for safety (for example, "investigations" become "learning reviews", "incidents" are "events", "findings" are "learning", "non-compliance" is "variation" etc.)
- **Change the approach**: introducing new processes and tools & techniques
- **Change the outcome**: a longer term objective to improve safety, system performance and human well-being

Features of the programme are that it is:

- Additive – to augment existing H&S systems
- A person-centred approach
- Aiming to redefine safety as the presence of positives, not the absence of negatives
- Focus on understanding decision making rather than punishing non-compliance
- Ultimately to improve performance through enhanced learning

Adam introduced the concept of "Restorative Just Culture" focussed on restoring trust, confidence and everyone's well-being following an incident, rather than seeking someone to blame. With this, traditional safety investigations are replaced with "Learning Reviews" where attention is given to "sensemaking" – understanding the point of view and perspective of the people involved and the thinking around the actions they took.

Thirty years of learning by accident

Graham Braithwaite, Director of Transport Systems and Professor of safety and accident investigation at Cranfield University reflected on the last three decades of system safety, which also coincided with his career in safety. Graham said that when he started out as a student, the UK transport sector was still reeling from major disasters such as the Herald of Free Enterprise at Zeebrugge, British Midland at Kegworth, and the Clapham Junction and Purley rail accidents. On one level, such events galvanised resolve to prevent such events from recurring, but on the other, it showed that there was also a level of risk acceptance that thankfully is no longer the case.

Graham went on to discuss the progress that has been made and said that the accident record demonstrates that a huge amount has been achieved. He said the role of "systems thinking" has been highly significant, not least in helping to understand the complexities of accident causation and the myriad influences on safety performance, both good and bad.

Milestone accidents have been major drivers for improved safety, but this reactive approach has also been increasingly supplemented by more proactive systems thinking, and safety science has evolved as our understanding of technical failures, human performance and organisational influences have advanced. Many new systems have been introduced, such as proximity warning systems that have improved safety, but they come at a cost of increasing complexity and opacity where operators, who are called upon more rarely in critical situations, have even less experience of dealing with them.

Graham also mentioned how the pandemic has had a complex effect on safety management, for example, maintaining recency of flight experience of pilots and even the infestation of insects in pitot tubes in grounded aircraft.

Milestone accidents which have taken place over that period remind us however, that progress has often come at the price of lives lost and property or environmental damage.

Graham covered the challenges of future flight and the race to meet the environmental challenges for aviation, not only in developing new fuel technologies but also in ensuring that novel solutions remain safe. For example, air mobility solutions have myriad issues with regards to new regulations, technology, operational, air traffic management and social acceptability. Graham said that there is now unprecedented change underway with many new disruptive technologies and many new players with novel societal impacts.

Graham concluded that the safety community has never been more important both now and in the future.

"Cowboy digital" undermines safety-critical systems

Prof. Harold Thimbleby, See Change Fellow in Digital Health, based at the University of Swansea discussed the prevalence of what he called "Digital Cowboy" programmers, who produce software that has little consideration to the safety implications of failure.

He illustrated this first by showing a poorly wired extension lead that he was PAT testing, which although appearing to work well, contained a hidden and potentially life-threatening fault.

His next example was an RCD tester that had a programmed but undocumented (invisible) feature that had resulted in readings that were erroneous and hence unsafe, for example, it registered a reading for 50V on a live circuit that was actually closer to 240V. As with the previous example, the system appears to be working when it fact it is in an unsafe state.

Another example he gave was a high-end domestic coffee machine that was physically well-made but had "Cowboy software engineering" as evidenced from its user interface; with confusing placement of a decimal point in a temperature reading to indicate which of the machine's two boilers was being measured.

Harold finally presented an example from the Healthcare domain where a system was being showcased for perinatal care but carried a disclaimer that it was not intended for diagnosis or treatment decisions. Harold said that this would not be permitted in Aviation systems like jet engines, but it is widely tolerated elsewhere as seen from the previous examples. Harold concluded with a final call that as an industry "we must do better".

At the interface of engineering safety and cyber security

Dr Reuben McDonald is Head of System Safety, Security & Interoperability at High Speed Two (HS2) Ltd, which is building a new high speed railway connecting towns in the South, Midlands and North of England. Reuben said that a modern railway is a highly electrotechnical system with connectivity and networks inherent in its design. For High Speed 2 this means a design which is based on a confluence of different wired and wireless networks. These networks support the delivery of safety critical, safety related, operation critical and wider business functions.

His talk explained how HS2 is assessing its safety and cyber risks in an efficient manner, leveraging products and assessments from the application of the Common Safety Method on Risk Assessment to support cyber assessments based around IEC 62443 risk methodology and associated control.

What do Byzantine Generals and Airbus airliners have in common?

Dewi Daniels from Software Safety Ltd, first described the events surrounding a China Airline's flight involving an Airbus A330 in 2020. During the routine landing, all three primary flight control computers shutdown causing several systems to become unresponsive, such as the spoiler deployment, reverse thrust and auto-brake. Both pilots were then forced to manually brake the aircraft, which eventually stopped just short of the end of the runway.

Dewi went on to explain the architecture of the flight control systems where each of the flight control computers has an independent (and diversely developed) monitor that can shut down the computer if processing outputs disagree.

In the Air China incident, all three computers failed and Dewi drew parallels between this failure and the "Byzantine General Problem". In the Byzantine General Problem, a General in charge of several Lieutenants must get agreement on common actions amongst the lieutenants for the overall campaign to be successful.

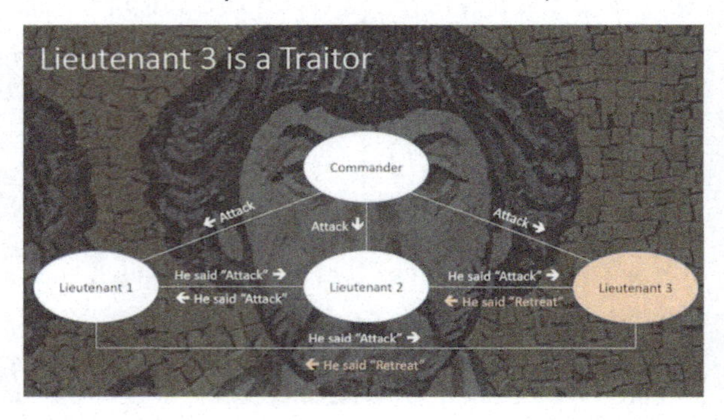

However, one of the commanders is unreliable and so may send misleading messages to the other commanders to sabotage the campaign. Dewi said that the well-known computer scientist, Leslie Lamport, wrote a paper in 1982 describing a solution to this form of problem where each Lieutenant cross-checks the General's instructions with each other to discover the unreliable Lieutenant.

Dewi then likened this to a 2 out of 3 voting system where the same input is given to 3 separate processors and results from each compared. Failures of one of the processing units can then be detected by the other two, which would share a common output value.

While this gives added resilience for individual processing unit failures, it does not help when the inputs are erroneous to start with, and this is what happened in the Air China incident – Weight on Wheels signal fluctuation, the pilot moving the rudder pedal and clock drift, meant that the monitors calculated different values from the main processors and shutdown all computers within the space of a second.

Thought for the Symposium

Tim Kelly from the Church of England and former Managing Director of the SCSC, concluded the event with his "Thought for the Symposium". You can find the full transcript of his talk on page 37.

Banquet

Despite reduced numbers, the traditional evening banquet was a huge success with not one, but two, speakers during the evening. Wendy Owen gave an excellent pre-dinner speech where she discussed her involvement and experiences with co-authoring the "30 Years of Safety Systems" book. Wendy's talk is reproduced in full on page 33.

We then had our first ever virtual after-dinner talk from Tom Anderson, former Managing Director of the SCSC. Tom's entertaining talk discussed the lessons he'd learnt on how to give after-dinner talks after listening to 30 years of after-dinner speakers. Such lessons were: "Brevity is the soul of wit" and quoting Tom Lehrer: "Plagiarise, plagiarise, let no one else's work evade your eyes, but please call it research!"

On a more serious note, he shared the economist Tim Harford's concerns around "Motivated Reasoning" (similar to "Policy-based Evidence" and "Confirmation Bias") – finding facts and seeking experts that support your claim.

Tom concluded with reference to one of the quiz questions in which he had featured – asking whether he was famous for reciting the Robert Burns poem "Tam O'Shanter". In lieu of reciting all 200 lines of the poem, Tom gave an amusing summary of the poem, and with its final message to be wary of the costs incurred "Whene'er to Drink you are inclin'd", he toasted the Pearl anniversary of the club with a glass of Scotch Whisky!

Report by Paul Hampton, SCSC Newsletter Editor

SCSC

FOR EVERYONE WORKING IN SYSTEM SAFETY

Bookings at:
www.scsc.uk/events

This seminar is an opportunity to hear about how coding and software engineering for safety-critical systems will change in the future.

It will be useful for software engineers, safety practitioners and for those involved in the planning and management of critical software projects.

Details at: www.scsc.uk

How Coding will Change for Safety Systems

www.scsc.uk

THE SAFETY-CRITICAL SYSTEMS CLUB

Seminar: The Future of Coding for Safety-Critical Systems

9th June 2022, Wellcome Collection, London and online

This seminar examines the future of coding for safety systems: what languages and techniques are likely to be used, how model-based approaches may be incorporated, what coding standards and guidelines may be applicable, the agile methodologies employed, and how verification and analysis can be integrated with the process of developing code.

The speakers will cover a variety of topics related to the area. They are all experts in their field, and can be relied upon to give informative and engaging talks:

Rod Chapman, Protean Code - "Rust" (TBC)

Les Hatton, Oakwood Computing Associates - (TBC)

Andrew Banks, MISRA - "The C language... you wouldn't start from here!"

Paul Sherwood, Codethink - "Safety for software-intensive systems"

Nick Tudor, D-Risq - TBA

There will be a discussion session at the end where delegates are encouraged to raise further questions with the speakers.

This event will be held at the Wellcome Collection, 183 Euston Rd, London NW1 2BE

Wendy Owen's SSS'22 Pre-dinner speech

Wendy Owen, currently a Research Associate in fusion energy at Bangor University, gave a pre-dinner speech at the SSS'22 Banquet, providing insights into her own career in system safety and describing the highs and lows of co-authoring the "30 Years of Safer Systems" anthology of SCSC Newsletter articles, published to celebrate the 30th Anniversary of the Safety-Critical Systems Club.

So, here we are, 30 years after the SCSC came into being. That's about half a lifetime ago for some attendees, including myself, and – thankfully perhaps – some of us are still around as well.

The SCSC came into being at a significant time in my own life. Back in 1991, I was in the last year of UKAEA's graduate scheme, working at Winfrith in the wilds of Dorset, conducting safety analyses of the UK's first civil small modular reactor (SMR) – what civil SMR you say? Yes, there was one in the pipeline back then, but the project was canned a year later. The idea and prospect of civil SMRs has actually only been resurrected very recently, and you may have seen news articles on the subject. I suspect, if climate change had been taken more seriously back then, we might have a few SMR power stations by now. That year, I also found myself conducting reliability analysis on another first, as far as I know – the first computer-based trip system for a UK nuclear plant, at Dungeness B. I also had an unforgettable work trip to Switzerland (long story, ask me later).

The following year I was relocated to the Safety & Reliability Directorate (SRD) in Cheshire. It was whilst at SRD that I first became aware of the SCSC, and subsequently signed up on-and-off over the years. At some point, I even wrote a Newsletter article called "The Right Ballpark", which was about data uncertainties and risk bandings, in an era before the risk matrix approach became popular.

> **One of my previous colleagues once made an astute observation that, "Wendy is at her best with 'Projects In Distress' "**

30 years later, I find myself back in the nuclear sector – this time in the fusion arena – having worked in many other sectors since, and also unexpectedly working in academia. This means I find myself with a bit more time on my hands for non-work type work compared to when I was in industry, including supporting organisations like the SCSC.

This turned out to be rather convenient for Mike, as although he already had several editors primed for the 30th anniversary book, another one was needed if SCSC was going to get it out the door before this conference.

One of my previous colleagues once made an astute observation that "Wendy is at her best with 'Projects In Distress' ", which sometimes appears as a quote on my CV. I wasn't sure if Mike's email was going to result in a distress call. It turned out that one of the editors wasn't very well – indeed it later transpired that they were expecting a baby – and one of the book's chapters was called "Mayday Mayday Mayday". So, of course, I said Yes.

To put the timeframe in perspective, that call was last May, and we've been having almost weekly meetings since then, and there are five editors involved. That's a lot of work, and indeed, generally a lot of teamwork, as everyone assumed slightly different roles. Also sometimes a lot of debate – the healthy variety – especially when the original title was "30 at 30"; but then one day, we realised that over 40 papers were included. We tried to delete some, but honestly, it was really hard to delete any of them because they were all so good. So, instead, we changed the title to what you see today, and the book contains 46 papers, which, co-incidentally, is the number of conference participants in this room this evening. Anyway, there's always a way around a problem, I like to think.

Within the book, there are the "green themes", or chapter headings, which have – by any engineer's admission – some great titles, mainly thanks to Louise Harney:

- AS LOW AS RUMOURED POSSIBLE
- AYE, ROBOT
- COMMUNICATION FAILURE
- CULTURE VULTURES
- DESIGN FOR UNSAFETY
- FROM HARDWARE TO HELL
- MAYDAY MAYDAY MAYDAY
- BLACK SWANS
- RUBBER–STAMPED
- RUN DON'T WALK
- SIGN YOUR LIFE AWAY HERE
- DRIVER, PLEASE STEP OUT OF THE VEHICLE
- THE FIRST STEP IS ACCEPTANCE
- THE FORGOTTEN ELEMENT

And lastly, TWO ARPS TO SOLVE THEM ALL

(I still can't remember what an ARP is, but it sounds like you'd need one).

My own contribution was mainly to the "blue boxes", which involved finding notable accidents and significant published legislation, standards or guidelines for specific years, or other technology-related milestones, and writing some poignant text about them, one for each newsletter article. However, they couldn't be any old accident or standard, because Louise had already drafted the chapter introductions – those themed "green boxes" – and the newsletter articles had already been selected too. The blue boxes hence needed to act as a link between Louise's green boxes and the original newsletter articles. It all needed to connect, make sense and be a good read. Suffice to say, I lost count of the number of blue boxes I'd written, but the rest of the team liked them, so that was OK!

Mike, Paul and Roger were conducting the same exercise, so we also had to make the blue boxes look consistent, as if they were written by the same person – for bonus points you can try to guess who wrote which blue box!

The others wanted to bring a few non-safety events into the picture; I wasn't convinced about this approach and ended up clamping down on specific terrorism events, but was unable to stop the flow of advancing mobile phone technology and certain infamous contemporary artworks, so please don't blame me for those...

I recall 1999 and 2000 produced a lot of text due to the Y2K issues, and there were sometimes so many railway and aviation accidents it all got a bit confusing which year was which, and which to make case studies of. Occasionally, for some years, there simply weren't enough highlights, which was actually harder to deal with as we scraped around trying to find interesting things to talk about.

Sometimes years and articles got accidently muddled up – no mean feat to sort out when you're working with a 300-page document, but Paul came to the rescue on many editorial occasions. Paul, I have to say, is also an expert document formatter (if he wasn't before).

For those of you with a penchant for numeric data, I expect you to ponder for a while over the penultimate page, page 304, which provides a Year Index and also cross-references to papers from the 25th anniversary edition. In this 30th edition, 1998, 2010 and 2017 ended up as the most prolific years, with four papers each, whilst 1995, 1997 and 2011 didn't make the grade, with no representative papers. We thought we'd point this fact out now because otherwise some of you would no doubt pick up on it, and perhaps even complain about it! However, we don't have an answer as to why 1995, 1997 and 2011 turned into "gap years", but you will find that the 25th edition also had gap years (in fact more than this current edition). Hence, we propose this as a challenge for whoever writes the 35th edition!

Whilst all this was going on, we made attempts to contact all the original authors of the articles, to check that they were happy for their articles to be published – and also, if possible, to provide a postscript. Many were tracked down through personal contacts and LinkedIn (I think I found a calling as a detective), and in fact, most authors have provided postscripts, for which we are all very grateful.

Many postscripts are technical in nature, some are very short, others are large essays, but I have two favourite postscripts myself – a one-liner essentially saying something like "nothing much has changed in 110 years or so since this article was written", and another, that alludes to the fact that the author, since writing the article, changed from caffeinated to decaf coffee, and nowadays, thinks quite differently. I am sure you will find your own favourites hidden in there too. You will also find many highlighted quotes from the articles scattered throughout the book, where we ended up with blank pages. I think these add a nice touch that my old O-Level English Literature teacher would have highly approved of before he abandoned our class and ran off with a Jazz band; consequently, I didn't get a particularly good grade for O-Level Literature and have spent a long time trying to compensate for it, including this speech.

I have two favourite postscripts myself – [one] essentially saying something like "nothing much has changed in 110 years or so since this article was written"

Back to the book. We all reviewed each other's editorial work and contributions, and the document went through two main drafts before the publication issue.

As with any good independent safety review process, suffice to say, we ended up with quite an extensive Corrective Actions sheet. With a 300-page book, let us tell you that it's a lot easier to spot problems on a paper version you can flick through rather than on a computer screen, so it was actually available on Amazon a lot earlier than you think it was, because we had to buy the paper drafts... Fortunately, I had accumulated a few Amazon vouchers the previous year and these were put to good use.

To finish, it has definitely been an enjoyable, and enlightening exercise, and we hope that you will appreciate everything that we've written, and what the article authors have written, both past and present.

Also please don't forget to read the Final Words from the Editors on the final page, which I will read for you here this evening for convenience:

And so it is that, with a virtual nod and a bow to all these enlightening papers and their erstwhile authors, we bid our farewells to the last three decades of safer systems.

As we edit this book in the midst of a global pandemic due to COVID–19, we know that this sort of event has been predicted for many years, and yet it appears we have been caught largely unprepared. Moreover, many of the systems – from infection spread models to vaccination certificate apps – used to manage the virus have been put together quickly and, in many cases, have been found wanting. This is a real source of frustration; we could have taken a much more engineering and assurance-led approach to management of the epidemic if we had been prepared ...

We can only contemplate, from our own COVID hidey-holes, with ongoing advances in technology and increasing concerns about climate change, what challenges the next decade will bring, one earlier-than-expected fine spring morning, basking in the increasingly hotter summer sunshine, hidden amongst the autumnal leaf scatterings or thrown up in the midst of more severe winter storms. Will it be autonomous self-driving vehicles? Will it be planes and trains running on hydrogen? Will it be regular rocket travel to Mars?

As safety engineers, we very much hope that we make a difference to people's lives, and it is true to say that we do save people's lives, in our own way. We may not be NHS paramedics, firefighters, the RNLI or Mountain Rescue, but we do make the systems, upon which these organisations depend, safer.

Paul, Louise, Wendy, Mike and Roger. December 2021

Until next time, I hope that this book will grace many coffee tables (and I live in hope that maybe next time someone will pick up on "The Right Ballpark"). Thank you for listening, and please... keep safe!

Wendy Owen

Wendy Owen is currently a Research Associate at Bangor University. In 2021 she moved into academia, following a 30+ years career mainly in engineering consultancy, managing safety & dependability cases, projects, bids and teams, and covering a wide range of highly-regulated industries including nuclear, defence, rail, process, aviation and automotive. She currently works in the Safety, Regulation & Policy Group within the Nuclear Futures Institute (NFI) at Bangor University. She supports development of design safety guidelines for future fusion power plant, covering safety, radiological safety, security and environmental protection.

See the back cover for further details of the "30 Years of Safer Systems" book.

Thought for the Symposium

Tim Kelly, former Professor of High Integrity Systems at the University of York and former Managing Director of the SCSC, provides his closing remarks in the form of a "Thought for the Symposium", to conclude proceedings at the 30th Anniversary Safety-Critical Systems Club Symposium.

Firstly, I'd like to say that Mike and the team are to be congratulated at what they have been able to achieve in running this year's symposium as a blended conference – bringing together folks from near and far, on-site and online, together once more to share ideas, learn from one another and hopefully – by doing so – advance the field and practice in industry. The importance of bringing people and ideas together should never be underestimated, and it's what I'd like to focus on for the next few minutes.

I read this week some words from my new boss – Justin Welby – the Archbishop of Canterbury (not my ultimate boss, you understand!) some ideas that challenge the narrative that has (at points) emerged during the pandemic – and for this symposium it has made me reflect on some of the problems that we encounter in the field of safety engineering.

Justin Welby said that "One person put it best when they said it was as though the pandemic had caused us all to 'lose the muscle-memory of how to be together'". Covid had shown unequivocally that individualism and atomisation were both illusion and fantasy; from staying at home to bulk buying supplies, getting the vaccine, or wearing a face mask, the message was clear: "Our actions affect other people. We cannot do what we want without it having an impact somewhere else." It struck me that these ideas may have something to say to us in the world of safety engineering.

Anyone who has been on one of my Goal Structuring Notation (GSN) courses, will have heard me talk about "divide and conquer" as one of the strategies to tackle large complex problems and break them down into smaller more manageable problems. It's a natural thing for us as engineers to do. It's what we do! Some would say that it's the true sign of an engineer that perhaps even from an early age they enjoyed taking something apart into its constituent bits and then putting them all back together again!

You could say that this is even what a symposium such as this one does: it takes the complex phenomena of system safety and breaks it down into separate sessions, separate insights and papers on data safety, or autonomy, or human factors, and so on. However, when I used to teach "divide and conquer" in my GSN courses, the most alert attendees would, of course, say "But hang on ... what about the interactions? What about how one sub-problem cannot be truly solved without impact another?" And of course they were right, and we'd then talk of ways to cover the interconnections and interactions, perhaps with a part of the safety argument specifically ring-fenced to address that concern.

If you're honest – you breathe a sigh of relief when a particular deviation is proposed and you realise that this particular issue is not your problem

But what does this interconnectedness of all things look like in the middle of our safety engineering activities, on a day-to-day basis. Well, perhaps you can relate to the following observations: have you ever found yourself in the middle of hazard analysis ... a HAZOP or an FHA, where the list of potential concerns and problems to resolve seem to be piling up, and – if you're honest – you breathe a sigh of relief when a particular deviation is proposed and you realise that this particular issue is not your problem – it lies outside the boundary of your concern. Phew – that one, at least, is someone else's problem – or you're under no obligation to fix it given the contract you've signed up to. That would be a maintenance issue, you might say, and we're not responsible for maintenance, or that would be a human factors concern, but that's out of scope. Of course, it doesn't necessarily mean it isn't a problem, but to put in terms of the pandemic – "not my toilet roll!"

Or maybe you've found yourself drawing up the issues to be addressed by your safety or assurance case, and again if you're honest, you're relieved when some tricky claim, or some challenging part of the argument can be declared outside the scope. One project team that I was involved in that was using Modular GSN to construct what turned out to be quite a complex argument, seemed to *really enjoy* and *relish* using "Away Goals" (those claims that you've realised are a necessary part of the argument, but are addressed elsewhere) as a way of declaring something *they* didn't have to do. The metaphorical picture of a gardener throwing snails over the garden fence comes to mind, or perhaps those areas marked on ancient maritime maps which simply said, "Here be dragons" at the edge of the explored territory.

Of course, this issue of system safety being a holistic – whole – system issue isn't a new observation, but highlighting the relevance of holistic thinking I was struck by a newspaper article just this last week provocatively titled: "Covid lockdowns did more harm than good". The article reported on the results of a controversial study that had concluded that the costs to society far outweighed the benefits and called for lockdown to be questioned as a future pandemic policy.

The study reported that some lockdown measures may have increased deaths by stopping access to outdoor space, "pushing people to meet at less safe places", while isolating infected people indoors, where they could pass the virus on to family members and housemates, and of course, challenging people's mental health.

It was an interesting (if albeit since contested) study. In terms of what we have been discussing within this symposium, it served to focus attention of the dangers of ever simply fixating on one problem, and potentially one mitigation (one solution, if you like) at the detriment of many other problems. One of the things that makes me recognise this problem most keenly at a personal level at the moment is that one of my friends – a fellow vicar – is now in a hospice, undergoing palliative care for advanced and aggressive cancer that remained unobserved, and undiagnosed for the vast majority of 2020 and 2021, in large part, because she simply could not get face-to-face access to her local GP, which was operating in a highly restrictive mode because of the lockdown measures. Her case, unfortunately, is not isolated. Cancer doctors and researchers are currently experiencing a bow wave of undiagnosed cancer cases because of the pandemic.

System safety engineering can sometimes seem like a game of "whack-a-mole"

During the pandemic we experienced at first hand the emphasis on one set of statistics, one set of indices and charts – of "flattening the curve" – in the daily number 10 briefings. We are perhaps only now starting to see the impact of the many 'curves' that weren't flattened, or of the other indices that were rising. A reminder, if ever we needed one, of the danger of focusing on single set of metrics.

And talking of "flattening the curve", it brings to mind how the activity of system safety engineering can sometimes seem like a game of "whack-a-mole". I'm sure you've seen this fairground game, where you're given a big hammer with which you are to bash on the head of a mole as soon as it appears ... only as soon as you knock one mole down another pops up to be knocked down ... and then another and so on.

One of the earliest group hazard analysis exercises we used to teach at the University of York on the Safety Critical Systems Engineering MSc was called the "Aircraft Configuration Check". In that system, the maintenance engineer had to write down a part number for a faulty unit, walk it over to the stores, retrieve the replacement part and then enter the details of the new part back into the configuration management unit. Lots of potential for error – that was the idea! However, I was always struck how some – in definite "fix it" mode – would simply suggest that if the whole process could be managed by computer and networked connections, then the problem would be solved.

I would then point out that this was a problem *changed*, rather than solved. As soon as one problem was eliminated, another potential problem would be created. Whack-a-mole, see? It doesn't mean that we wouldn't make the change, but we should never be so naive as to think that apparent progress doesn't bring with it its own set of contingent problems.

So what are the takeaways for us today? My observations over the recent years of the symposium are that, firstly, given the complexity of systems we are increasingly proposing and integrating into everyday life, we will never run out of challenges and problems to address in this field!

I hope you've all had a chance to read the article that Paul Hampton has put together for the 30th Anniversary Edition of the SCSC Newsletter entitled, "The Future of System Safety". To collate ideas for the article, Paul asked contributors to imagine it was SSS'52 and we were celebrating the 60th anniversary of the club. He asked us to imagine possible titles for the keynotes and the challenges that we would be facing: For my part, my contribution was a keynote on "Regulating Safety in the Metaverse" and a session on "The Safety of Healthcare Nano-bots". I invite you to take a look at the rest of the article!

However, my other observation is that we don't even need to look into the future to see that the sub-disciplines and aspects of safety engineering are getting more and more interconnected and entangled (and rightly so). Even reviewing the papers and programme of this year's symposium, highlights how, in many regards, it's increasingly difficult to "break off a chunk" of a problem without it impacting on other aspects – whether that be the interaction of safety and consumerism that we heard of from John McDermid on day one, or the necessity of consideration of ethics that we heard from Paula, or the need to manage dependencies in Human Factors in Rachel's talks yesterday, or Gary's talk today of how the march of progress in multi-core processors provides challenges for safety. One person's progress is another person's headache. One person's cost-efficient solution ("we don't need LIDAR for autonomous vehicles", I hear Elon Musk say), is at the expense of another person's software complexity, and yet another's assurance claims to be addressed.

John Donne once famously wrote in his poem, "No Man is an Island"

> *No man is an island entire of itself; every man is a piece of the continent, a part of the main; if a clod be washed away by the sea, Europe is the less, as well as if a promontory were, as well as any manner of thy friends or of thine own were; any man's death diminishes me, because I am involved in mankind. And therefore never send to know for whom the bell tolls; it tolls for thee.*

(Apologies for the non-inclusive 17th Century language). As we survey the rich panoply of safety engineering and its many disciplines, as a symposium such as this allows, we might well, from whatever vantage point we take (whether that be data safety, or processor safety, or organisational safety, or any other), heed John Donne's words that challenge isolationism.

Rather than gazing from afar and perhaps thinking "phew, I'm glad that's not my problem", we should remember "ask not for whom the bell tolls, it tolls for thee!" The discipline is increasingly interconnected and interdependent. If anything, as those reports on the pandemic and the effects of lockdown have shown, the need for thinking more holistically about our safety problems will only increase in time and heighten in importance.

As well as John Donne, I'd like to end by suggesting that the words of Jesus may give us a particular perspective on the problem. The Bible records that a rich man once asked him what the most important commandments were to follow. After telling him that the most important commandment was to love God, Jesus told him that the second most important commandment was similar, but was to "love others as you love yourself" ... a commandment that highlights the interconnectedness that exists amongst us all. So, if there's a final thought for me to end the symposium with it would perhaps be this, "to love other safety engineers as much as you love yourself", and alongside this "to care for other's safety problems and challenges as much as you care for your own."

Thank you.

Tim Kelly

Tim Kelly worked for over 25 years in the domain of high-integrity and safety-critical systems engineering and was Managing Director of the SCSC from 2016 to 2019. Over the years he has published many seminal papers in the field of systems safety and developed the Goal Structuring Notation (GSN). In 2019 he took a 'leap of faith' and gave up being a full-time Professor in High Integrity Systems at the University of York to become a vicar in the Church of England at Beverley Minster.

Bletchley Park Tech Trip

Saturday 30th April 2022 saw the inaugural SCSC Tech Trip, which was to Bletchley Park, once the top-secret home of the World War Two Codebreakers. Fourteen attended including SCSC members and their families. Dewi Daniels provides a report of the event.

I last visited Bletchley Park about 20 years ago. The site has seen a lot of restoration work carried out since then. The last time I visited, many of the buildings were dilapidated and very few of them were open to the public. This time round, it was great to see the site rejuvenated and the ground and buildings well-maintained.

I arrived early, so I was able to enjoy a delicious bacon bap in the Bletchley Park Coffee Shop before I met the others at 1130. The first hour was free to wander around the site. It's quite a large site and very pleasant to stroll around on a sunny day. There are many interesting places to visit, including Alan Turing's office in Hut 8, which is where they broke the German naval ciphers.

At 1230, we had an hour-long guided tour of the site. Our excellent guide said that his link to Bletchley Park was that he had worked for "a branch of the Foreign Office". We learned about the history of the site from its purchase by Admiral Sir Hugh Sinclair, head of MI6, in 1938 to its closure in 1946, including an explanation as to how the hut system worked.

We then had an hour for lunch in one of the cafes at Bletchley Park, after which we moved to the National Museum of Computing.

Although the National Museum of Computing is in Block H on the wider Bletchley Park site, it is outside the gates of the Bletchley Park Museum. Whereas the Bletchley Park Museum seems very well-funded and commercially aware, with an excellent gift shop and several places to eat, the National Museum of Computing seems to be run by enthusiasts, with lots of interesting kit packed into a small space.

We'd booked a 2-hour guided tour, which turned into 3 hours. I think the guide appreciated our clear interest and the many questions that we asked. We saw many amazing exhibits, such as a replica Bombe and Colossus. The original Bombes were built to break the Enigma code.

Colossus was the world's first programmable electronic digital computer, built to break the Lorenz code that was used by the German High Command. The replica Bombe and Colossus took an incredible amount of effort to build. The replica Bombe took 13 years, the replica Colossus took 15 years.

We were also shown the Harwell Dekatron, which is the World's oldest working digital computer. It's a strange device. It's decimal rather than binary and is very slow (about 5 seconds to carry out a multiplication), but apparently very reliable.

We were also shown a replica of the EDSAC, which was the second electronic digital stored-program computer (the first was the Manchester Mark 1), several mainframes including an ICL 2966, many early personal computers, several early laptops and personal organisers. I found the National Museum of Computing surprisingly interesting. I was astonished at how many of the computers are still in working order.

It's a shame that these two fine museums don't cooperate more closely. Bletchley Park is well-funded, the ground and buildings are well-maintained, but most of the buildings are empty. On the other hand, the National Museum of Computing next door is full of interesting computers, most of them in working order, but is very short of space.

All in all, it was a very enjoyable day out. I look forward to the next SCSC Tech Trip!

The next SCSC Tech Trip will be to the Royal Air Force Museum Cosford, Shropshire on 1st October 2022 and includes a guided tour.

This is a museum dedicated to the history of aviation and the Royal Air Force in particular. There are thrilling displays of aircraft including the world's oldest Spitfire as well as other attractions such as iconic cars, models and tanks.

The Tech Trips are open to all – family and friends are more than welcome!

Accident Investigation and Safety Culture Event Report

The 'Accident Investigation and Safety Culture' webinar was held online on the 27th April 2022 by the Safety Culture Working Group. The event opened with Mike Parsons welcoming delegates and gratefully thanking our speakers. Michael Wright, Chair of the Safety Culture Working Group, reports on proceedings.

The event had several themes, including:

- Does learning from accidents support the improvement of safety performance?
- How can you encourage learning from incidents and accidents?

These themes were set in the context of "Safety II" that addresses criticisms of what is characterised as "Safety I" including that:

"... seeing deficiencies in hindsight does nothing to explain the generation or persistence of those deficiencies." [1]

"I would be delighted if Root Cause Analysis would disappear, but I am not very optimistic. The simplicity of the method and the thinking behind it is too attractive to be overcome by sound arguments against its practical value."

Erik Hollnagel, Safety Management Trend Report 2017.

Overcoming a culture of safety silence to prevent accidents

Catherine Baker, Director of the Confidential Incident Reporting and Analysis Service (CIRAS) kicked off with an excellent talk on why people can remain silent about safety concerns and how this may be overcome, especially through confidential reporting. Catherine commenced with a musical reference to Simon and Garfunkel's "The Sound of Silence", particularly the lines:

- People talking without speaking
- People hearing without listening
- And no one dared
- Disturb the sound of silence
- Silence like a cancer grows
- Hear my words that I might teach you

A very engaging reference and poignant link to her talk.

Catherine went on to cite some incidents where someone knew that something could happen, including visible erosion that led to a train derailment, a ship taking a turn too fast and a worker habitually breaching safety rules. In all cases there was "endemic silence". This was built upon by a film of a worksite at which a person fell from scaffold, where 66 people either fail to notice or mention the visible unsafe conditions. Catherine presented an anatomy of silence, shown below. In this model, if someone notices and speaks up, this helps recognise and act on unsafe conditions. Clearly if someone fails to speak up, an unsafe condition may persist. Catherine drew out many factors that can contribute to silence, such as fear of adverse reactions, large power-distance relationships, and a sense of individual vulnerability.

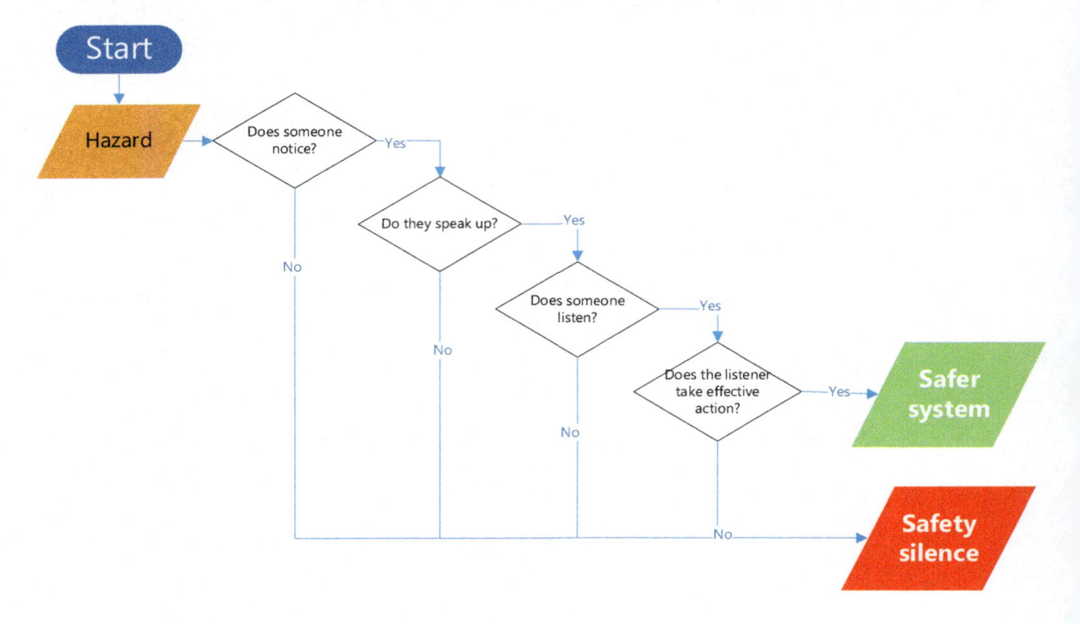

Catherine highlighted several methods for overcoming a culture of silence, including:

1. Overcoming fear of:

- Reprisals
- Peer reaction
- Being incorrect
- Not being listened to
- Damaging relationships

2. Closed loop guarantees:

- I get to see what action is taken
- I can challenge back if the action has missed the mark

CIRAS provides one means by which these fears can be overcome. It provides a confidential reporting process. Catherine was also confidently able to cite many examples where CIRAS had linked reports back to the responsible organisations, leading to safety improvements. This recognises the importance of people being able to see that action is taken in response to their concerns.

Discussion explored several points, including why organisational cultures may create endemic silence and why there is a need for confidentiality. It was also discussed whether CIRAS is a reactive approach to safety improvement – the answer being that by reporting concerns early on, this prompts improvements that prevent the accident from occurring.

Nuclear Safety Culture: From the biggest accident that never was (Davis-Besse) to the realities of building the first new nuclear plant in a generation

Tom Hughes (Nuclear Safety Culture Lead at Hinkley Point C) talked us through the 'biggest accident that never was' – at a plant called Davis-Besse (shown on the right here) in the USA in 2002. Severe corrosion was discovered on the reactor pressure vessel that put the plant in grave danger. Minor leakage of coolant corroded the carbon steel head. Over time, a cavity formed around the control rod. The cavity was discovered when someone touched the rod and it leaned over. This condition had been developing over years and the warning signs were there. The first warning of potential corrosion was issued in 1988, a leak was spotted in 1991, with further leaks in 1992-96, none of which were adequately investigated. Concerns were raised about inadequate cleaning in 2000.

Reflecting on the lessons learned from this near-miss, the talk considered the challenges surrounding developing and embedding a robust safety culture on the largest construction site in Europe, Hinkley Point C. The challenges include:

- Construction mindset
- Complex contracting model
- Enabling senior leadership engagement
- Large number and high "churn" in staff

A successful response to these challenges entailed a multi-faceted strategy illustrated in the figure below. Key aspects include: encouraging learning about nuclear safety, influencing the supply chain, improving awareness of the link between safety and quality and pushing the importance of safety from the top down.

The strategy includes clear elements of reporting and sharing learning. Again, this highlights the importance of facilitating an open and learning culture.

Discussion highlighted how EDF has engaged the supply chain, many of whom had no nuclear safety experience, in the principles of nuclear safety.

Organisational influences on safety: RFA Argus – Wildcat Case Study

Cdr Steve Gamble provided a gripping summary of a series of aviation accidents. By using a case study from a Wildcat helicopter incident in 2017, he used this 'near-miss' to provoke debate and highlight some of the long term organisational and human behaviours which may adversely influence safe operations.

The helicopter attempted to take off while only one of two sets of engines were engaged. The crew recognised the problem and landed back on the edge of the deck.

This occurred during an extended 14-day training period, where multiple crews performed flights under supervision of instructors, including night flights.

The investigation determined that there was a higher than standard ratio of students to instructors, and use of less experienced crew than the norm. At the time of the incident it was thought that crew were fatigued, and distracted by events such as a delayed launch. There were five missed opportunities during take-off where the error could have been spotted by the crew. The design of the rotor controls did not meet good human factors, with the potential to overlook that the rotor engine settings were not both at the same position.

Concerns had been raised about the ratio of instructors to students in previous incidents. It was thought that the high ratio of students to instructors had become normalised. The discussion explored why lessons from previous incidents had not been carried forward, and why known shortcomings in training arrangements were tolerated. It was speculated that lessons learned had not been carried forward, possibly due to churn in personnel and the passage of time.

Just culture, psychological safety & facilitating learning from error

Michael Wright presented the findings from a review of research completed for the Energy Institute [2] into Just Culture and the concept of psychological safety. In contrast to Catherine Baker's talk, this one explored how to encourage openness within an organisation and overcome barriers to speaking up. The importance of an open culture was highlighted by the conclusions of the inquiry into the 2005 Texas City refinery explosion that said:

> *"BP should involve the relevant stakeholders to develop a positive, trusting, and open process safety culture within each U.S. refinery ... establish a climate in which: workers are encouraged to ask challenging questions without fear of reprisal..."*

On a more positive note, a report from RSSB (2018) was quoted that said that over the last 15 years, a 90% reduction in SPADs (Signals Passed at Danger) has been seen in the rail industry and they attribute it, in part, to an "open and mature safety culture".

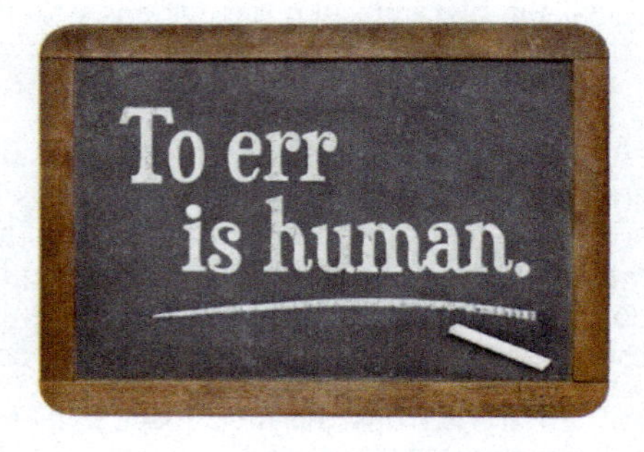

The presentation went on to first define Just Culture as *"An atmosphere of trust in which people are encouraged (even rewarded) for providing essential safety-related information, but in which they are also clear about where the line must be drawn between acceptable and unacceptable behaviour."* [3] It went on to cite research that reported a mixed impact of implementing Just Culture. While some studies cite increases in reporting of errors and near misses and decreases in the number of respondents who reported feeling fearful of reporting, others reported little evidence of an upward trend in non-punitive responses to error and little evidence of improvement in safety performance. This was thought to be related to a fear of adverse consequences, especially for 'blame worthy' vs 'blameless' acts.

It was noted that some businesses had reformulated the 1997 version of the Just Culture culpability decision flow chart, to focus on the role of organisational culture in violations and the individual's intent. In practice, most violations are prompted by an "organisational optimisation" motive and action should focus on the organisation rather than blaming the individual.

The talk went on to define psychological safety as "A shared belief held by members of a team that the team is safe for interpersonal risk taking" [4] and contrasted it with Just Culture. It drew out some parallels, such as both aiming to learn from error. Psychological safety was said to focus more so on individual and team factors that may inhibit people, such as being a new employee, and how psychological safety may be facilitated by actions such as:

- Team building (developing trusting interpersonal relationships)
- Having (inclusive) leaders and role models facilitating learning through adoption of a set of supportive behaviours, accessibility, neutral language and positive reinforcement
- Providing assurance of no adverse personal consequences from reporting error
- Demonstrating the value of speaking up by acting effectively on feedback and reporting actions back to people

Error needs to be seen by the team and the organisation as a learning opportunity; and a shared experience about what works and what does not work. Learning from error is a collective responsibility aimed at performance improvement.

The concept of an error management culture was cited, quoting Guchait et al. (2014) [5]. Error management culture *"involves organizational practices related to communicating about errors, sharing error knowledge, quickly detecting and handling errors, and helping in error situations."*

The discussion explored the importance of an open culture in recognising and acting on safety concerns but also in helping to solicit new ideas to improve safety performance, especially in hierarchical organisations with high power distance relationships.

Learning from recent major accidents

There was a robust discussion about the Boeing 737 MAX and VSS Enterprise crashes facilitated by Michael Wright and Mike Parsons, focusing on the role of safety culture in these incidents and the value of learning from events. In both cases, single point failures with potential catastrophic consequences were tolerated within the design.

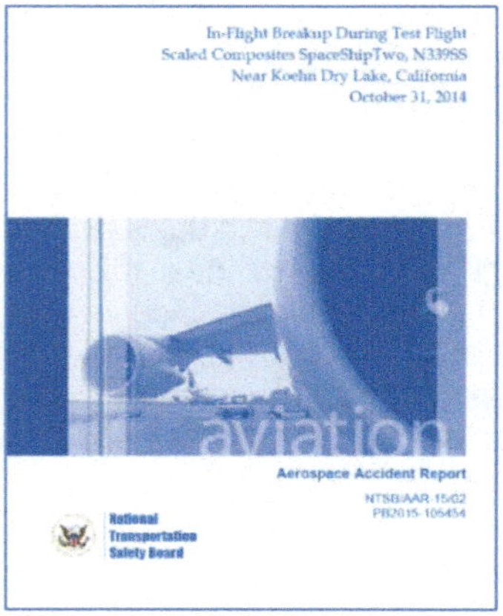
In-Flight Breakup During Test Flight
Scaled Composites SpaceShipTwo, N339SS
Near Koehn Dry Lake, California
October 31, 2014

aviation

Aerospace Accident Report

NTSB/AAR-15/02
PB2015-105454

National
Transportation
Safety Board

VSS Enterprise was the first SpaceShipTwo (SS2) spaceplane built for Virgin Galactic; it had a Feather flap assembly with twin tailbooms positioned upwards to stabilise attitude and increase drag on re-entry. These needed to be deployed (during ascent) by the time Mach 1.4 speed was achieved, so that the crew could abort before reaching Mach 1.8, thereby avoiding excessive speed on re-entry in the event that the feather mechanism did not deploy. This would give the crew a few seconds to be sure they could go supersonic.

The feather mechanism was deployed too early while still under rocket propulsion at Mach 0.8. The feathering mechanism then began moving due to aerodynamic forces on the tail and inertial loads, and the craft disintegrated. It was noted that the pilot had about 30 seconds to do three tasks from memory, while experiencing high acceleration and vibration which had not been replicated in simulator runs and of which they had no recent experience. There was no call and response system between the two pilots.

The Hazard analysis had not considered the possibility of error of premature unlocking of the feather mechanism, and it had been assumed that the pilots would always operate correctly due to their training and simulation runs. Despite failing to meet hazard analysis requirements, the Federal Aviation Administration waived the hazard analysis. The SS2 preapplication process began about 2 years before the developer submitted its initial application but after the vehicle had been designed and manufactured. The NTSB concluded that *"there was 'a lot of pressure, political pressure' to issue experimental permits, even when FAA/AST evaluators were uncomfortable with an application, which diminished AST's safety culture."*

The well-known Boeing 737 MAX accidents involved the automatic nose-down trim commanded by the Maneuvering Characteristics Augmentation System (MCAS) forcing two planes into dives. The accident investigation reported that:

- The FAA were largely unaware of an automated flight-control system that played a role in the crashes
- The FAA management overruled the determination of the FAA's own technical experts at the behest of Boeing
- Boeing employees cavalierly dismissed the FAA
- Boeing concealed crucial information from the FAA and pilots
- There was knowing reliance on a single sensor and automated activation despite reported concerns

It was also reported that Boeing had a business objective for the 737 MAX to not require any simulator training for pilots who were already flying the 737 NG.

It was suggested that in both cases, the development and regulation was impacted by a range of factors as shown in the figure below.

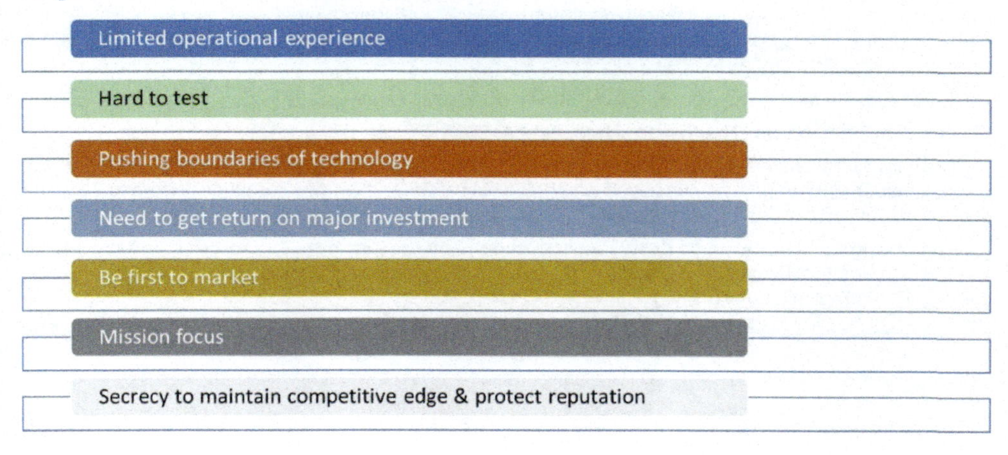

The discussion explored three questions:

1. What are the pressures on safety climate from developing new (expensive) technology?
2. How do you create an organisational climate that assures safety and adherence to recognised standards in the context of experimental and new technology?
3. Is there value in learning from accidents?

Delegates stated that it was clearly essential to learn from accidents. Indeed, it was noted that a different response to the first Boeing 737 MAX accident may have helped to prevent the second crash.

The discussion also explored how good safety culture can be eroded, with the suggestion that safety culture is fragile and may be adversely impacted by influences such as production and competitive demands.

As regards maintaining a safety culture during the development process, the role of a truly independent and effective regulator was highlighted. As stated by the NTSB chairman Christopher Hart in relation to the Virgin SS2 crash: *"Many of the safety issues that we will hear about today arose not from the novelty of a space launch test flight, but from human factors that were already known elsewhere in transportation".* This also highlighted the value of sharing good practice and lessons learned between sectors.

The discussion concluded that it is unclear how a business may mitigate pressures on its culture and avoid erosion of a good safety culture. This was thought to be a good topic for a future event!

References

[1] Prof E.Hollnagel, Prof Wears and Prof Braithwaite. From Safety-I to Safety-II: A White Paper. 2015. https://www.england.nhs.uk/signuptosafety/wp-content/uploads/sites/16/2015/10/safety-1-safety-2-whte-papr.pdf Accessed May 2022

[2] Michael Wright and Samuel Opiah. Literature review: The relationship between psychological safety, human performance and HSE performance. 2018. https://publishing.energyinst.org/heartsandminds/research Accessed May 2022

[3] J. Reason. Managing the risks of organizational accidents. Ashgate, 1997.

[4] Prof Amy Edmondson. The fearless Organisation. 2018

[5] Guchait, P, Paşamehmetoğlu, A, and Dawson, M. Perceived supervisor and co-worker support for error management: Impact on perceived psychological safety and service recovery performance. August 2014. International Journal of Hospitality Management 41:28–37.

[6] The Boeing 737 MAX Aircraft: Costs, Consequences, and Lessons from its Design, Development, and Certification Preliminary Investigative Findings, https://transportation.house.gov/imo/media/doc/TI%20Preliminary%20Investigative%20Findings%20Boeing%20737%20MAX%20March%202020.pdf Accessed May 2022

Connect

The Newsletter and eJournal

Do you have a topic you'd like to share with the systems safety community? Perhaps an interesting area of research or project work you've been involved in, some new developments you'd like to share, or perhaps you would simply like to express your views and opinions of current issues and events. There are now two publishing vehicles for content – shorter, more informal content, can be published in the Newsletter with longer, more technical peer-reviewed material more suitable for the eJournal. If you are interested in submitting content, then get in touch with Paul Hampton for Newsletter articles: paul.hampton@scsc.uk or John Spriggs for eJournal papers: john.spriggs@scsc.uk

The SCSC Website

Visit the Club's website thescsc.org for more details of the Safety-Critical Systems Club including past newsletters, details of how to get involved in working groups and joining information for the various forthcoming events.

Facebook

Follow the Safety-Critical Systems Club on its very own Facebook page.

www.facebook.com/SafetyClubUK

Twitter

Follow the Safety-Critical Systems Club's Twitter feed for brief updates on the club and events: @SafetyClubUK

LinkedIn

You can find the club on LinkedIn. Search for the Safety-Critical Systems Club or use the following link:

www.linkedin.com/groups/3752227

Advertising

Do you have a product, service or event you would like to advertise in the Newsletter? The SCSC Newsletter can reach out to over 1,000 members involved in Systems Safety and so is the perfect medium for engaging with the community. For prices and further details, please get in touch with the Newsletter Editor.

SCSC Working Groups

The Safety-Critical Systems Club is committed to supporting the activities of working groups for areas of special interest to club members. The purpose of these groups is to share industry best practice, establish suitable work and research programmes, develop industry guidance documents and influence the development of standards.

Assurance Cases

The Assurance Cases Working Group (ACWG) has been established to provide guidance on all aspects of assurance cases including construction, review and maintenance. The ACWG will:

- Be broader than safety, and will address interaction and conflict between related topics
- Address aspects such as proportionality, rationale behind the guidance, focus on risk, confidence and conformance
- Consider the role of the counter-argument and evidence and the treatment of potential bias in arguments

In Aug 2021, the group published v1.0 of the Assurance Case Guidance: scsc.uk/scsc-159

One of the working group's activities is the maintenance of the Goal Structuring Notation (GSN) Community standard.

See scsc.uk/gsn for further details.

In May 2021, the group published v3.0 of the standard: scsc.uk/scsc-141C

Lead **Phil Williams** phil.williams@scsc.uk

SCSC Working Groups

Security Informed Safety

The Security Informed Safety Working Group (SISWG) aims to capture cross-domain best practice to help engineers find the 'wood through the trees' with all the different security standards, their implication and integration with safety design principles to aid the design and protection of secure safety-critical systems and systems with a safety implication.

The working group aims to produce clear and current guidance on methods to design and protect safety-related and safety-critical systems in a way that reflects prevailing and emerging best practice.

The guidance will allow safety, security and other stakeholders to navigate the different security standards, understand their applicability and their integration with safety principles, and ultimately aid the design and protection of secure safety-related and safety-critical systems.

Lead Stephen Bull stephen.bull@scsc.uk

Data Safety Initiative

Data in safety-related systems is not sufficiently addressed in current safety management practices and standards.

It is acknowledged that data has been a contributing factor in several incidents and accidents to date, including events related to the handling of Covid-19 data. There are clear business and societal benefits, in terms of reduced harm, reduced commercial liabilities and improved business efficiencies, in investigating and addressing outstanding challenges related to safety of data.

The Data Safety Initiative Working Group (DSIWG) aims to have clear guidance on how data (as distinct from the software and hardware) should be managed in a safety-related context, which will reflect emerging best practice.

An update to the guidance (v3.4) was published in Jan 2022: scsc.uk/scsc-127G

Lead Mike Parsons mike.parsons@scsc.uk

SCSC Working Groups

Safety of Autonomous Systems

The specific safety challenges of autonomous systems and the technologies that enable autonomy are not adequately addressed by current safety management practices and standards.

It is clear that autonomous systems can introduce many new paths to accidents, and that autonomous system technologies may not be practical to analyse adequately using accepted current practice. Whilst there are differences in detail, and standards, between domains many of the underlying challenges appear similar and it is likely that common approaches to core problems will prove possible.

The Safety of Autonomous Systems Working Group (SASWG) aims to produce clear guidance on how autonomous systems and autonomy technologies should be managed in a safety-related context, in a way that reflects emerging best practice.

The group published v3 of its guidance Safety Assurance Objectives for Autonomous Systems, in Jan 2022 scsc.uk/scsc-153B

Lead **Philippa Ryan** pmrc@adelard.com

Multi- and Manycore Safety

It is becoming harder and harder to source single-core devices and there is a growing need for increased processing capability with a smaller physical footprint in all applications. Devices with multiple cores can perform many processes at once, meaning it is difficult to establish (with sufficient evidence) whether or not these processes can be relied upon for safety-related purposes.

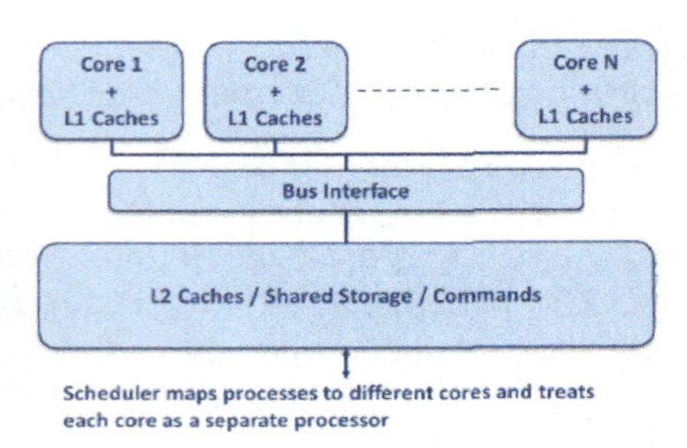

Parallel processes need to access the same shared resources, including memory, cache and external interfaces, so they may contend for the same resources. Resource contention is a source of interference which can prevent or disrupt completion of the processes, meaning it is difficult to know with a defined uncertainty the maximum time each process will take to complete (Worst Case Execution Time, WCET) or whether the data stored in shared memory has been altered by other processes.

The Multi- and Manycore Safety Working Group (MCWG) has been established to explore the future ways of assuring the safety of multi- and manycore implementations.

Lead **Lee Jacques** Lee.Jacques@leonardocompany.com

SCSC Working Groups

Ontology

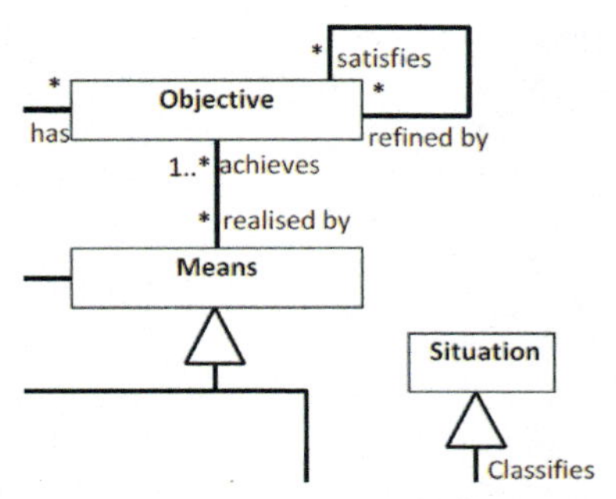

The Ontology Working Group (OWG) develops ontologies that will form the basis of SCSC guidance, as well as having wider industrial and academic applications.

The OWG is currently working on the definition of an ontology of risk for application in guidance for risk-based decision making – notably safety and security – and for which ISO 31000 Risk Management principles are to be applied.

The Data Safety Working Group (DSIWG) developed the core aspects of the Risk Ontology, which has been migrated to this working group. The Risk Ontology will form the upper ontology to the Data Safety Ontology that the DSIWG will continue to develop.

Lead Dave Banham ontology@scsc.uk

Covid-19

The Covid-19 Working Group is involved with discussion, analysis and assistance related to the Coronavirus. The group meets remotely to see what a systems and assurance view of the situation brings.

The group has compiled an extensive range of Covid-19 related material and made this available on the working group's website pages along with ongoing developments in the thoughts and ideas of the group.

Members are all experienced engineers, used to making reasoned arguments about safety. The aim is to apply the groups considerable technical expertise to the problem and find and assure appropriate solutions.

Lead Peter Ladkin ladkin@causalis.com

SCSC Working Groups

Service Assurance

Risks presented by safety-related services are rarely explicitly recognised or addressed in current safety management practices, guidelines and standards. It is likely that service (as distinct from system) failures have led to safety incidents and accidents, but this has not always been recognised. The Service Assurance Working Group (SAWG) has been set up to produce clear and practical guidance on how services should be managed in a safety-related context, to reflect emerging best practice.

The group published v3.0 of the guidance in Jan 2022: scsc.uk/scsc-156B

Lead Mike Parsons mike.parsons@scsc.uk

SCSC Safety Culture

The Safety Culture Working Group (SCWG) has been established to provide guidance on creating and maintaining an effective safety culture. The group seeks to improve safety culture in safety-critical organisations focussed on product and functional safety, by sharing examples and latest approaches collated from real-life case studies.

Meetings provide an opportunity to discuss any particular aspects attendees are interested in taking forward, and to help set future directions for the group.

The group's held a webinar on the 27th April 2022 called 'Accident Investigation and Safety Culture'. A report of the event is provided on page 45.

Lead Michael Wright michael.wright@greenstreet.co.uk

60 Seconds with ... Zoe Garstang

Zoe Garstang ARAeS is an Airworthiness Engineer at BAE Systems. She previously undertook an Advanced Engineering Apprenticeship with the company before joining the Airworthiness team in March 2020.

Zoe is also the lead for SCSC Safety Futures Initiative (SFI) looking to develop the next generation of safety engineers and works closely with the SCSC Steering Group to develop the club's Equality, Diversity and Inclusion strategy.

What first attracted you to working in the field of System Safety?

For me, the opportunity and awareness of System Safety came along (ironically) by accident. If it had not been for my apprenticeship Skills Coach, who encouraged me to try a placement in Airworthiness, I would probably not have gained the same appreciation for it. Whilst in Airworthiness, my placement manager mentioned the SCSC and asked if I would be interested in getting involved – and here I am!

What aspect of your career are you most proud of?

Up to now, I think this would be completing my apprenticeship. When I joined BAE Systems in 2016, I had been so torn between the apprenticeship or continuing onto college and university. I sometimes wondered if I was 'missing out' on something by going down a different route to my school friends. But that couldn't have been further from the truth. I was incredibly fortunate that my apprenticeship became so much more than just a day job.

How have you enjoyed leading the Safety Futures Initiative?

I have thoroughly enjoyed it. I've been able to pull on my experience in other volunteering roles whilst developing the SFI, and I've learnt a lot about the SCSC and its members along the way. It's great to see the club being so supportive of the initiative and I hope it can develop into a useful network for the next generation of Safety Professionals.

What future changes would you like to see in the field of System Safety?

The biggest change I would like to see is less technical but more to do with the interest and engagement with the next generation; not only ensuring we retain upcoming safety engineers, but also encouraging others to explore System Safety from a much earlier stage than they may currently be exposed to it.

What's your most favourite quote or motto?

My favourite quote comes from a film called **"**We Bought A Zoo**":**

> "Sometimes all you need is twenty seconds of insane courage ... And I promise you, something great will come of it".

I can be my own worst enemy when it comes to overthinking things at times, so it's often a case of plucking up the courage to take the first step.

If you could learn to do anything, what would it be?

Perhaps not one of the safest sports, but I have always wanted to learn to play ice or roller hockey. I've skated in both forms since I was young, but never taken it any further. Maybe I need 20 seconds of insane courage!

Sometimes all you need is twenty seconds of insane courage!

If you could be any fictional character, who would you choose?

I would really like to have the analytical skills of Sherlock Holmes. I learn a lot by observing others, but the way he can observe and deduce a situation is something else.

What's the best piece of advice you've ever been given?

An engineer I met on a work experience day, who ended up igniting my interest in engineering, always emphasised the idea of **"**building a 3D/spherical knowledge bank**"**; the idea being to go beyond your specialism, department, business function etc. and learn about how it all interconnects and builds. Ultimately, try to get as rounded an understanding about something as possible.

Which work of art or fiction inspired you the most when you were growing up?

I always loved Michael Morpurgo's books growing up, particularly *Born To Run*. The message of giving someone (people or animals) a chance and just taking the first step to building trust with them, and where that can lead, really stuck with me. It's also ironic to mention it now, as I've just adopted an ex-racing greyhound!

SCSC Membership

The SCSC provides a range of services to the System Safety community including seminars, tutorials, leadership events, specialist topic working groups, the annual symposium and a comprehensive body of publications. Membership brings many valuable benefits such as free access to online events, the SCSC Newsletter and access to presentations and other resources from events.

Individual Membership

To become an individual member of the SCSC please register on the SCSC website using the icon at the top right of any page and select "Register". Complete and save your account registration and then verify your email address. Once registered and logged in click the link "why not join the SCSC..." inviting you to become a member at the top right of the page or select "Pay membership" from the icon.

Individual membership can be paid online using a credit/debit card through our secure payment partner Realex Global Payments or contact Alex King for other payment methods. For student or retired member rates please contact Alex King to get your account status changed.

Corporate Membership

Your company contact with the SCSC should arrange the membership and any renewals for your organisation. To join as a member covered by a corporate membership, register as per the instructions for an individual member and then contact Alex King to confirm your affiliation.

Renewing Membership

You should be notified by email when your membership is almost expired or shortly after it has expired. These notifications will contain a link to the online renewal page or you will be able to renew when logging onto the website through the 'click to renew' link.

Membership Fees

The following fees are applicable for new and renewing members:

- 1 year Individual Membership: £125
- 2 year Membership: 20% discount: £200
- 3 year Membership: 33% discount: £250 (3 years for the price of 2)
- 1 year SFI Membership: FREE for first year, £35 for years 2 & 3
- 1 year Membership, retired member rate: £35
- For Corporate Membership discounts contact Alex King.

A one-month Publication Pass is also available for £15. This allows access to all SCSC website publications in a particular calendar month.

Contact Alex King using office@scsc.uk

The SCSC Steering Group

Tom Anderson

Honorary member

Stephen Bull

stephen.bull@scsc.uk

Jane Fenn

jane.fenn@scsc.uk

Paul Hampton

paul.hampton@scsc.uk

James Inge

james.inge@scsc.uk

Nikita Johnson

nikita.johnson@scsc.uk

Tim Kelly

Honorary member

Mark Nicholson

mark.nicholson@scsc.uk

Mike Parsons

mike.parsons@scsc.uk

Roger Rivett

roger.rivett@scsc.uk

Emma Taylor

Honorary member

Sean White

sean.white@scsc.uk

Robin Bloomfield

Honorary member

Dewi Daniels

dewi.daniels@scsc.uk

Zoe Garstang

zoe.garstang@scsc.uk

Louise Harney

louise.harney@scsc.uk

Brian Jepson

brian.jepson@scsc.uk

Graham Jolliffe

Honorary member

Alex King

alex.king@scsc.uk

Wendy Owen

wendy.owen@scsc.uk

Felix Redmill

Honorary member

John Spriggs

john.spriggs@scsc.uk

Phil Williams

phil.williams@scsc.uk

Club Positions

The current and previous (marked in italics) holders of club positions are as follows:

Managing Director

Mike Parsons 2019-

Tim Kelly 2016-2019

Tom Anderson 1991-2016

Steering Group Chair

Roger Rivett 2019-

Graham Jolliffe 2014-2019

Brian Jepson 2007-2014

Bob Malcolm 1991-2007

Programme & Events Coordinator

Mike Parsons 2014-

Chris Dale 2008-2014

Felix Redmill 1991-2008

Manager

Alex King 2019-

Newsletter Editor

Paul Hampton 2019-

Katrina Attwood 2016-2019

Felix Redmill 1991-2016

University of York Coordinator

Mark Nicholson 2019-

eJournal Editor

John Spriggs 2021-

Administrator

Alex King 2016-

Joan Atkinson 1991-2016

Website Editor

Brian Jepson 2004-

Safety Futures Initiative Lead

Zoe Garstang 2019-

Nikita Johnson 2019-2021

Calendar

February '22

M	T	W	T	F	S	S
	1	2	3	4	5	6
7	8	9	10	11	12	13
14	15	16	17	18	19	20
21	22	23	24	25	26	27
28						

March '22

M	T	W	T	F	S	S
	1	2	3	4	5	6
7	8	9	10	11	12	13
14	15	16	17	18	19	20
21	22	23	24	25	26	27
28	29	30	31			

April '22

M	T	W	T	F	S	S
				1	2	3
4	5	6	7	8	9	10
11	12	13	14	15	16	17
18	19	20	21	22	23	24
25	26	27	28	29	30	

May '22

M	T	W	T	F	S	S
						1
2	3	4	5	6	7	8
9	10	11	12	13	14	15
16	17	18	19	20	21	22
23	24	25	26	27	28	29
30	31					

June '22

M	T	W	T	F	S	S
		1	2	3	4	5
6	7	8	9	10	11	12
13	14	15	16	17	18	19
20	21	22	23	24	25	26
27	28	29	30			

July '22

M	T	W	T	F	S	S
				1	2	3
4	5	6	7	8	9	10
11	12	13	14	15	16	17
18	19	20	21	22	23	24
25	26	27	28	29	30	31

August '22

M	T	W	T	F	S	S
1	2	3	4	5	6	7
8	9	10	11	12	13	14
15	16	17	18	19	20	21
22	23	24	25	26	27	28
29	30	31				

September '22

M	T	W	T	F	S	S
			1	2	3	4
5	6	7	8	9	10	11
12	13	14	15	16	17	18
19	20	21	22	23	24	25
26	27	28	29	30		

October '22

M	T	W	T	F	S	S
					1	2
3	4	5	6	7	8	9
10	11	12	13	14	15	16
17	18	19	20	21	22	23
24	25	26	27	28	29	30
31						

November '22

M	T	W	T	F	S	S
	1	2	3	4	5	6
7	8	9	10	11	12	13
14	15	16	17	18	19	20
21	22	23	24	25	26	27
28	29	30				

December '22

M	T	W	T	F	S	S
			1	2	3	4
5	6	7	8	9	10	11
12	13	14	15	16	17	18
19	20	21	22	23	24	25
26	27	28	29	30	31	

January '23

M	T	W	T	F	S	S
						1
2	3	4	5	6	7	8
9	10	11	12	13	14	15
16	17	18	19	20	21	22
23	24	25	26	27	28	29
30	31					

Events Diary

26 May 2022 SCSC Seminar **Managing Unexpected Risks: Handling Rare and Severe Events Now and in the Future** **London, UK + Online** scsc.uk/e825	**1-2 June 2022** Conference **Reliability, Safety and Security of Railway Systems (RSSRail 2022)** **Paris, France** rssrail2022.univ-gustave-eiffel.fr	**9 June 2022** SCSC Seminar **The Future of Coding for Safety-Critical Systems** **London, UK + Online** scsc.uk/e912	**16 June 2022** SCSC Working Group **DSIWG#69: SCSC Data Safety Initiative Meeting #69** **Bath, UK + online** scsc.uk/e924
22 June 2022 SCSC Working Group **SAWG#42: SCSC Service Assurance Working Group Meeting #42** **Online** scsc.uk/e926	**5 September 2022** SCSC Working Group **SCWG: SCSC Safety Culture Working Group Meeting** **Location – TBA** scsc.uk/e921	**6 September 2022** Conference **SASSUR 2022: 9th International Workshop on Next Generation of System Assurance Approaches for Safety-Critical Systems** **Online** sites.google.com/view/sassur2022	**6-9 September 2022** Conference **41st International Conference on Computer Safety, Reliability and Security (SAFECOMP 2022)** **Munich, Germany** safecomp22.iks.fraunhofer.de
22 September 2022 SCSC Seminar **Seminar: Safety of Autonomy in Complex Environments** **London, UK + Online** scsc.uk/e890	**12-14 Oct 2022** Conference **RISK/SAFE 2022: 13th Conference on Risk Analysis, Hazard Mitigation and Safety and Security Engineering** **Rome, Italy** www.wessex.ac.uk/conferences/2022/risk-safe-2022	**22-23 Nov 2022** Conference **10th Scandinavian Conference on System and Software Safety** **Gothenberg, Sweden** safety.addalot.se/2022	**28 Nov 2022** SCSC Working Group **SCWG: SCSC Safety Culture Working Group (SCWG) Meeting** **Location – TBA** scsc.uk/e922

thescsc.org/membership

"30 Years of Safer Systems" contains articles from the last 3 decades of the Safety-Critical Systems Club (SCSC) newsletter "Safety Systems".

The book groups the articles into themes relevant to safety, with an introduction to the theme and a preface to each article giving major events from the year the article was first published, including accidents, incidents and positive improvements in safety. Themes include: Risk Assessment, ALARP, Artificial Intelligence/Machine Learning, Communication Failures, Safety Culture, 'Black Swan' events, Certification, Product Liability, Safety and Security Integration, Agile Methods, Data driven systems and Safety Cases.

Most of the original authors have provided a short postscript to their article to give extra context and explain progress in the intervening years.

Available for purchase on Amazon
www.amazon.co.uk/Years-Safer-Systems-safety-critical-Newsletter/dp/B09KNCYKDL

The Safety-Critical Systems Club Newsletter

Safety Systems

Vol 30 No. 3 - Oct 2022

THE SKY IS FALLING!

Assessing the risks from space weather

THERE BE DRAGONS!

Novel approaches to classifying data risks

MATRIX MANAGEMENT

Avoiding pitfalls when using risk matrices

For everyone working in Systems Safety

thescsc.org

SCSC Publication Number: SCSC-177

Contents

www.scsc.uk/sss

SSS' 23

Details of this and other events at: www.scsc.uk

Safety-Critical Systems Club Annual Symposium

www.scsc.uk

THE SAFETY-CRITICAL SYSTEMS CLUB

31st Safety-Critical Systems Symposium

7-9th February 2023, in-person in York and blended online

The Safety-Critical Systems Symposium in 2023 (SSS'23) will be held in York, UK and blended online as a live-streamed event. This event comprises three days of presented papers, including keynote presentations and submitted papers. There will also be a banquet on the Wednesday evening with additional talks.

This event will be run in-person at the Principal York hotel (located next door to York station) and also online through Zoom and the Whova application.

The Symposium is for all of those in the field of systems safety including engineers, managers, consultants, students, researchers and regulators. It offers wide-ranging coverage of current safety topics, focussed on industrial experience.

It includes recent developments in the field and progress reports from the SCSC Working Groups. It takes a cross-sector approach and includes the aerospace, automotive, defence, health, marine, nuclear and rail areas.

The symposium features are:

- Six keynote presentations and talks on submitted papers;
- Updates from the SCSC Working Groups;
- Audience participation and submitted questions.

And for in-person delegates in York:

- A banquet with after-dinner speaker on the Wednesday;
- Proceedings book and working group guidance books;
- Visit and demonstrations at the University of York Institute of Safe Autonomy;
- Special beer (or non-alcoholic alternative) designed and commissioned for the symposium;
- Safety Activities;
- Social events in York.

The symposium is a regular event and fosters a community spirit in the field: it is a great place to network, learn about the latest practice in safety and develop new business contacts.

For **further information, exhibition** and **booking queries** please contact: Alex King, Dept of Computer Science, University of York, Deramore Lane, York, YO10 5GH. Email: alex.king@scsc.uk

For **technical aspects** and **talks, speakers, abstracts, papers** or **posters,** please contact: mike.parsons@scsc.uk

Editorial

Last month the nation and indeed, the world, witnessed the passing of Queen Elizabeth II, ending an extraordinary life of public service – the UK's longest-serving monarch and one of the longest ever reigns in global history.

Her life covered almost a century of national and global affairs, spanning 15 UK prime ministers, and it's remarkable to think that while she leaves us now in a society where we have the beginnings of routine space flight, quantum computing and autonomous vehicles, she was born in an era before the electronic television was invented and decades before systems safety was, in any shape, formalised.

Her husband Prince Philip also enjoyed a long life, living to 99, but this might not have been the case when, aged 37, he came perilously close to a life-threatening hazard – the like of which we now try so diligently to avoid.

In 1958, he boarded an RAF Sikorsky Whirlwind helicopter travelling the short distance to RAF Patrington across the Humber Estuary. Mid-flight, the co-pilot spotted that the safety wire on the sliding cockpit window had snapped, and the window was moments away from coming off altogether and colliding with the helicopter's tail rotor, with possibly catastrophic results.

The co-pilot acted quickly by grabbing onto the window by bracing – and damaging – his knee to keep it in place as he battled against the aerodynamic forces. By now, over the Humber Estuary, and with no chance to land quickly, there was no choice but to carry on. In considerable discomfort, and with hands numb from the windchill, the co-pilot managed to keep hold as the helicopter made its way to safety.

In a sense this seems an apt metaphor for safety engineering in general. Sometimes we need to hold on doggedly to our safety assessments, designs, mitigations and indeed principles; even when this might be uncomfortable – both at an enterprise and personal level – but hold on we must; even till our hands are numb to avoid the murky waters of disaster that lie below.

In this edition of the newsletter there are three great feature articles, event reports from the "Managing Unexpected Events" and "The Future of Coding" SCSC seminars, an update from the Safety Futures Initiative and the "How do I get into Safety" series returns with insights from a few more members of the steering committee. Our 60 second interview is with Les Hatton who will be the after-dinner speaker at SSS'23.

It also gives me great pleasure to introduce some new features; we now have a "Recent Publications" section where new publications from the SCSC and other relevant sources will be listed and with thanks to Dewi Daniels we have our first ever book review!

To add in some of the fun that we enjoyed with the quiz at SSS'22, I am now including a System Safety crossword into the newsletter; so please try your hand at the questions! All correct answers received by the end of the year will be entered in the hat to win a £50 Amazon Voucher!

Paul Hampton
SCSC Newsletter Editor
paul.hampton@scsc.uk

In Brief

£270k fine after fatality involving mobile elevating work platform

A Lincolnshire based manufacturing company that specialises in lifting and handling equipment has been fined after an employee died after falling with a work platform onto the M25 motorway due to a miscalibration of the vehicle.

The miscalibration occurred through incorrect data being manually manipulated and uploaded onto the machine via a laptop. *hsmsearch.com*

NHS employees warned to expect three weeks of disruption as a result of a cyberattack

A hack on Advanced, a British software and managed services provider with 25,000 customers, has severely disrupted NHS IT systems and may prevent doctors from accessing patients' records for several weeks. *emcrc.co.uk*

KLM 737 used whole runway for take-off after intersection data slip-up

Pilots of a KLM Boeing 737-800 did not amend a runway intersection designation when recalculating take-off performance data for Amsterdam Schiphol, leading the aircraft to accelerate too slowly and use almost the entire runway length before becoming airborne. Dutch investigators have attributed the incident to operational and time pressures. *flightglobal.com*

UK to Have World's Longest Drone Superhighway

The UK government has announced it has given the go-ahead for the world's largest and longest network of drone superhighways to be built in the UK.

The drone superhighway will link cities and towns throughout the midlands to the southeast of the country, with the option to expand the corridor to any other locations in the country. *uasvision.com*

AAIP launch new guidance on the safety of autonomous systems

The Lloyd's Register Foundation-funded Assuring Autonomy International Programme (AAIP) at the University of York have published their latest guidance for the design and introduction of safe autonomous systems. The new research, "Safety Assurance of autonomous systems in Complex Environments (SACE)", marks an important milestone in providing a framework for the regulation and adoption of robotics and autonomous systems across a variety of sectors. *lrfoundation.org.uk*

Space Weather: Risks, Challenges and Solutions

Dr. Jonathan Eastwood discusses the risk of space weather, the evolving challenges it poses, and what is being done to meet those challenges. He explains what is meant by space weather and describes some of its main risk pathways and potential impact on human activity and technology both in space and on the ground. He then assesses the magnitude of the problem and concludes by highlighting some future trends around ensuring that the future of space is safe, secure and sustainable.

Space weather represents a significant threat to modern critical infrastructure and is an important socio-economic risk that has become increasingly prominent in the past two decades in the UK. It is arguably uniquely challenging, because although high-impact events are known to occur, their frequency is relatively low, meaning that society has not experienced a major space weather event in the modern technological era (e.g. the last 20 years). During this time, ever more complex technologies have transformed daily life, and we rely on space-based services in ever more intricate, and surprising, ways.

What causes space weather?

In the past 20 years, space weather has risen from being well known in the space science community to a subject that has caught the attention of the wider world to very high political levels [1]. In the UK specifically, space weather is listed on the national risk register [2], cited in the national space strategy [3] and is the subject of a dedicated policy paper on severe space weather preparedness [4].

Broadly speaking, the study of space weather is concerned with understanding how natural phenomena in space can have real and tangible impacts on human activity and technology both in space and on the ground. It is intimately related to the more general study of solar-terrestrial physics, which in itself is part of the field of heliophysics (The NASA Heliophysics Division [5] notes that it 'studies the nature of the Sun, and how it influences the very nature of space – and, in turn, the atmospheres of planets and the technology that exists there.')

The space weather phenomenon that most people are familiar with is probably the aurora, as illustrated on the previous page by a photograph taken in Alaska. Aurora are seen near the northern and southern poles, and have intrigued and inspired observers for thousands of years.

On occasion, tremendous auroral displays can occur, with vivid and bright features filling the sky. We now understand that these are caused by the Sun. Whilst the Sun appears somewhat featureless and uninteresting when seen at sunrise or sunset, in fact it is an incredibly active and dynamic object, with two main phenomena that interest us here: *solar flares* and *coronal mass ejections* (CMEs).

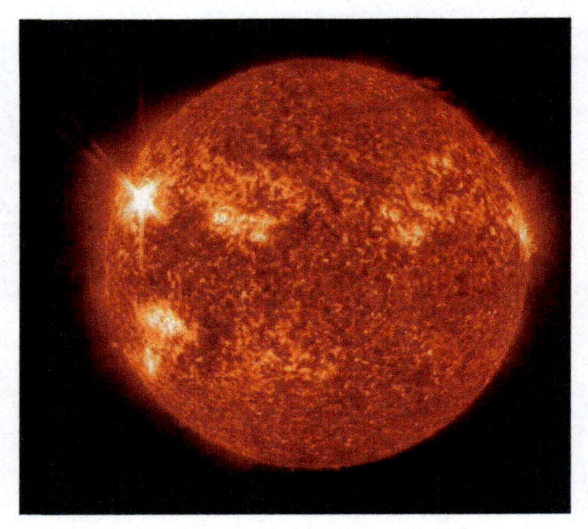

A solar flare is essentially a flash of light with emission across the electromagnetic spectrum, that lasts of the order of minutes. The image on the right was taken by NASA's Solar Dynamics Observatory and is a false-colour ultraviolet image of the Sun, designed to show the lowest layers of the solar atmosphere. The brighter regions are called *active regions*, which tend to be associated with more variability and stronger magnetic fields in sunspots. The flare is visible on the left-hand side, north of the equator. *Radio blackouts* are primarily caused by solar flares increasing ionisation of the upper atmosphere and distorting/disrupting signals.

A CME is an expulsion of material from the solar atmosphere travelling out through space at speeds of the order of 1000 km/s. The image below was taken by the ESA/NASA SoHO satellite and is a white light image of the solar atmosphere (the corona). The central blue circle blocks direct light from the Sun (outlined as a white circle), allowing the structure of the corona to be seen directly.

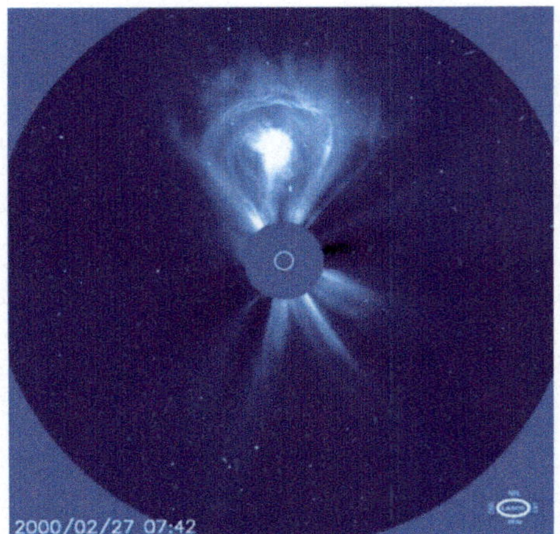

The Sun is the source of the solar wind, which is a supersonic stream of material (plasma) that is constantly emitted from the corona at speeds of hundreds of kilometres/second and fills the solar system.

The CME is the lightbulb shaped object at the top of the image, and a time-lapse movie created from many similar images can be used to trace the motion of the CME away from the Sun as part of the solar wind.

Whereas the light travel time to Earth is 8 minutes, and so flares are seen 'immediately', CMEs can take 2-3 days depending on their speed to reach the limit of Earth's orbit. CMEs travel radially from the Sun (more-or-less), and so establishing if a CME is Earth-directed is of critical importance for understanding the space weather risk.

If a CME arrives at Earth, and directly interacts with the Earth's magnetic field in space (the magnetosphere), it can drive a *geomagnetic storm* where energy and material circulates from the CME and solar wind into the magnetosphere. Through a chain of processes best described in terms of plasma physics, energy is stored in the magnetic field and explosively released, accelerating particles into the auroral regions and the Van Allen radiation belts. This causes the bright auroral displays referred to earlier. Of crucial importance is the magnetic structure of the CME. For certain magnetic field orientations, the interaction can be very weak, and little activity arises, but for others, a very strong interaction can occur. CME magnetic field structure can only be diagnosed by in situ measurements made by satellites just upstream of the Earth such as DSCOVR and ACE [6]. Thus, ground truth about whether a CME will cause a significant impact can only be ascertained shortly before they arrive at Earth. Whilst CMEs are the major source of storms, they can also be driven by structure in the background solar wind.

Finally, both solar flares and CMEs are also sources of so-called *Solar Energetic Particles* (SEPs), highly accelerated particles that represent a space weather risk in their own right, being the main source of *solar radiation storms*.

The impact of space weather

The following figure, produced by the European Space Agency, describes the wide variety of impacts that can arise as a consequence of space weather.

- **Space**: Astronaut radiation, satellite radiation damage, satellite solar cell degradation, satellite single event upset
- **Atmosphere**: Aircraft crew and passengers' radiation, HF radio wave disturbance, signal scintillation
- **Ground**: Decreased directional drilling accuracy, induced geoelectric field and current, geometrically induced currents in power systems

Of primary concern is the *loss of power*. Strong fluctuating currents in the ionosphere (essentially associated with the aurora) induce potential differences and electric fields in the Earth's surface, which can then couple into conducting infrastructure on the ground such as powerlines, pipelines and rail networks.

Of particular note here are the relatively famous events of 13 March 1989 in Quebec, where blackouts and transformer damage occurred as a result of a geomagnetic storm. The 2012 Royal Academy of Engineering report [7] suggests that 6 super grid transformers in England and Wales, and a further 7 in Scotland could be damaged and taken out of service as a consequence of severe space weather.

More recently, the Space Environment Impacts Expert Group's (SEIEG) 2022 Reasonable Worst Case Scenario [8] envisages that premature aging, and damage of transformers could occur, but also that transmission system voltage instability and voltage sag could occur, with tripping of safety systems potentially leading to cascade failure of the transmission network and/or regional outages.

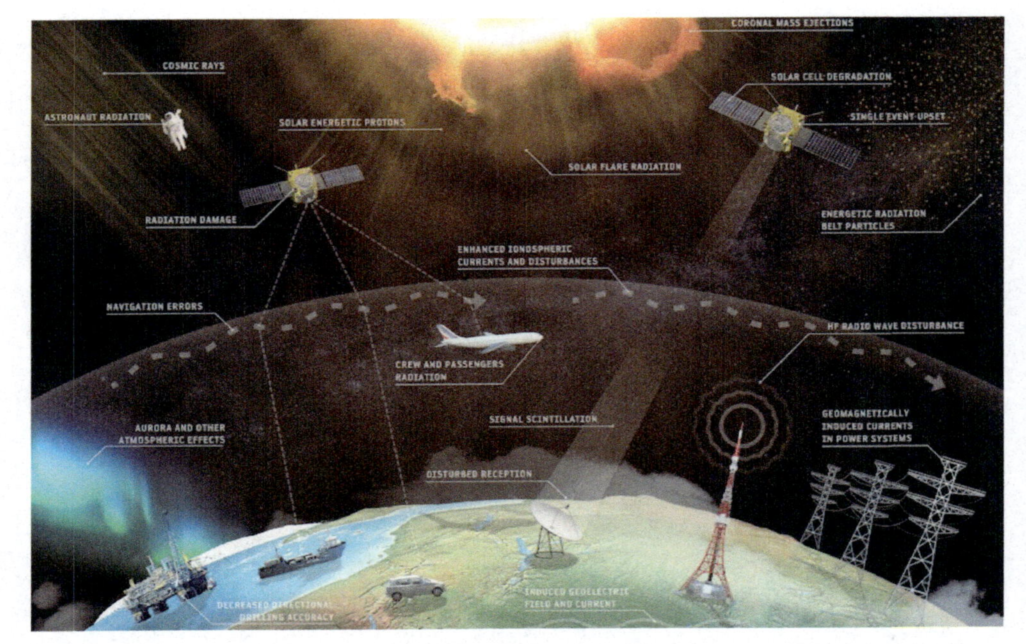

Another important area of impact is in Global Navigation Satellite Systems (GNSS) and position, navigation and timing (PNT) services (eg. GPS). GNSS can be disrupted by distortion to signals as a consequence of ionospheric disturbances driven by space weather. Outages could occur unpredictably and intermittently for several days during a major event, and the impact of this is still not well understood, particularly given the ever-increasing use of PNT services for a variety of downstream services.

Satellite anomalies have been extensively observed as a result of space weather. During what is arguably the last major period of space weather in October/November 2003, 33 Earth orbiting satellite anomalies were reported. Space weather can cause surface and deep dielectric charging, surface damage and solar panel degradation, background counts in sensors, and satellite drag and changes to orbit.

Outages could occur unpredictably and intermittently for several days during a major event, and the impact of this is still not well understood

These issues are thrown into stark relief by the rise of mega-constellations. With the aim of providing high-speed internet access anywhere on Earth, companies such as Starlink, One Web, Amazon Kuiper, Telesat and others offer the potential to transform daily life for billions of people. Starlink has already launched more than 2000 satellites into orbit, and if all planned mega-constellations come to fruition, tens of thousands of satellites will orbit the Earth in the next 10-20 years. Mega-constellations have already been impacted by space weather. In February 2022, a batch of Starlink satellites re-entered after launch because of increased atmospheric drag. In this case, a small geomagnetic storm heated the upper layers of the atmosphere, causing it to expand and increasing the (small) atmospheric density at the initial satellite orbit. This increased the drag on the satellites, slowing them down, and ultimately causing them to re-enter.

Finally I note that air travel is also subject to space weather considerations. For example, polar flights have been required to divert, because HF radio communications near the poles were disrupted by space weather. ICAO regulations now require space weather advisory information to be provided to the international aviation community.

Understanding, addressing and mitigating space weather risks

Space weather is a very difficult problem to assess. Firstly, the historical record shows that very severe events can happen, but they have not really happened in the modern era. A major event, such as experienced in 1859 (the so-called Carrington event) and with a return period of approximately 100 years, is expected to have a very significant impact on a wide variety of technological systems. Smaller events happen more frequently, and power-law behaviour is a feature of probability distributions describing various space weather phenomena.

It is highly likely that system failures will interact with each other to cause cascading failure modes that are fundamentally difficult to predict, and this is an area of research that has not received a great deal of attention

Secondly, each interval of severe space weather is unique, because it is ultimately driven by solar activity and so different combinations of flares, CMEs and solar energetic particles may occur depending on the underlying solar drivers.

The Sun's rotation period means that from the point of view of the Earth, an active region will take about two weeks to transit across the solar disk, and in a severe space weather period, one may expect a two-week interval of elevated risk, during which time several different types of phenomena could occur in unknown order, severity, etc. It is highly likely that system failures will interact with each other to cause cascading failure modes that are fundamentally difficult to predict, and this is an area of research that has not received a great deal of attention, in part because of the complexity involved.

A third challenge to understanding space weather is that it is very much impact-led and driven by a desire to reduce the socio-economic footprint. Therefore, the appropriate response is a calculation that is fundamentally risk based, considering the exposure, vulnerability and hazards. This calculation is also dynamic, in the sense that as some risk is retired (for example through engineering and design choice), new risks can be introduced (such as the dramatic increase in the use of PNT services, and the wider interest in e.g. driverless cars).

This is related to the question of space weather impacts on internal organisational activity and business resilience. Many organisations are not aware of their reliance on space infrastructure, or internal exposure to space weather issues. Similarly, organisations may not be aware of how their products and services may be exposed to space weather impacts. This exposure could arise at many points and could be of particular importance where certain levels of service are guaranteed.

Here, it is typically necessary for the analysis to be client-led, where the specific organisation enlists the help of expert groups or space weather service providers to understand what risks might be in play. It is important to recognise that space weather may well turn out not to be a major issue, but it is still necessary to investigate the possibilities as the unique nature of the space weather hazard means that surprising impacts may be identified.

The Future

At the national and international level, considerable efforts have been made to improve our ability to monitor and forecast space weather. In the UK, the Met Office is the 'owner' of the space weather risk, and operates the Met Office Space Weather Operations Centre (MOSWOC) [9] that provides 24/7 space weather forecasting. In the US, NOAA's Space Weather Prediction Center plays a similar role, and ESA's Space Safety office enables comparable space weather service provision.

Space weather forecasting relies on data and modelling, and whilst there are several satellites monitoring the Sun and the near-Earth space environment, there are plans to develop a new generation of observers, such as ESA's Vigil [10] space weather mission, that will provide new capabilities improving forecast quality and reliability. New generations of space weather models are being produced, in the UK the Space weather innovation, measurement, modelling and risk (SWIMMR) programme [11] funded by UK Research and Innovation is delivering new capabilities from the academic community into operational use at the Met Office.

Finally, space is a critical national infrastructure, and space weather is part of a wider drive to ensure the future use of space is safe, secure and sustainable. Work by the Space Lab network of excellence at Imperial College London has identified 5 key areas in this regard [12] – the terrestrial environmental impacts of space activities, space debris, planetary defence, space weather and space traffic management.

In particular, the impact of mega-constellations cannot be ignored, as they bring multiple issues about space safety, security and sustainability into sharp relief, raising many questions about policy and regulation.

Summary

Looking to the future, our use of space is transforming rapidly, and the benefits it brings has already changed many aspects of daily life. As organisations become ever more aware of how space technology is inextricably woven into the fabric of everyday life, the demand for knowledge about space weather is only likely to grow; where this touches on safety-critical systems, this may be particularly important. I hope this article will help raise awareness in the wider safety community about the risks of space weather, the challenges it poses, and what is being done to meet those challenges.

References

[1] The Economic Impact of Space Weather: Where Do We Stand? Eastwood et al., *Risk Analysis*, Volume 37, Issue 2, pg 206, 2017 https://dx.doi.org/10.1111/risa.12765

[2] UK National Risk Register 2020, https://www.gov.uk/government/publications/national-risk-register-2020, accessed Sept 2022

[3] UK National Space Strategy (Feb 2022), https://www.gov.uk/government/publications/national-space-strategy/national-space-strategy, accessed Sep 2022

[4] UK Severe Space Weather (Sep 2021) Preparedness Strategy, https://www.gov.uk/government/publications/uk-severe-space-weather-preparedness-strategy, accessed Sep 2022

[5] NASA Heliophysics Division, https://science.nasa.gov/heliophysics, accessed Sep 2022

[6] DSCOVR https://solarsystem.nasa.gov/missions/DSCOVR/in-depth and ACE https://solarsystem.nasa.gov/missions/ace/in-depth, accessed Sep 2022.

[7] Extreme space weather: impacts on engineered systems and infrastructure, https://raeng.org.uk/media/lz2fs5ql/space_weather_full_report_final.pdf, RAEng, 2013

[8] Summary of space weather worst-case environments (3rd revised edition), https://epubs.stfc.ac.uk/work/51273983, SEIEG, 2022

[9] MOSWOC, https://www.metoffice.gov.uk/weather/learn-about/space-weather/met-office-role, accessed Sep 2022

[10] Vigil, https://www.esa.int/Space_Safety/Vigil, accessed 2022

[11] Space weather innovation, measurement, modelling and risk (SWIMMR), https://www.ukri.org/what-we-offer/our-main-funds/strategic-priorities-fund/space-weather-innovation-measurement-modelling-and-risk, accessed Sep 2022

[12] Space Lab evidence-based policy reports, 2020, 2021 https://www.imperial.ac.uk/a-z-research/space-lab/research/

Image Attribution
Northern lights over the Poker Flat Research Range, Alaska USA 16/02/17 Credit: NASA/Terry Zaperach
Solar flare observed by NASA's Solar Dynamics Observatory 03/11/11 Credit: NASA
Coronal mass ejection observed by the LASCO coronagraph on ESA's SoHO Credit: SOHO ESA & NASA
ESA space weather effects graphic © ESA/Science Office CC BY-SA 3.0 IGO

Dr. Jonathan P. Eastwood, Reader in Space Physics, Imperial College London

Dr Jonathan Eastwood is a Reader in Space Physics in the Department of Physics' Space and Atmospheric Physics group at Imperial College London. His research interests include space weather, space plasma physics, and developing novel miniaturised instrumentation for space weather monitoring. He also serves as Director of the Space Lab Network of Excellence, which brings together all of Imperial's space capabilities under one network.

"30 Years of Safer Systems" contains articles from the last 3 decades of the Safety-Critical Systems Club (SCSC) newsletter "Safety Systems".

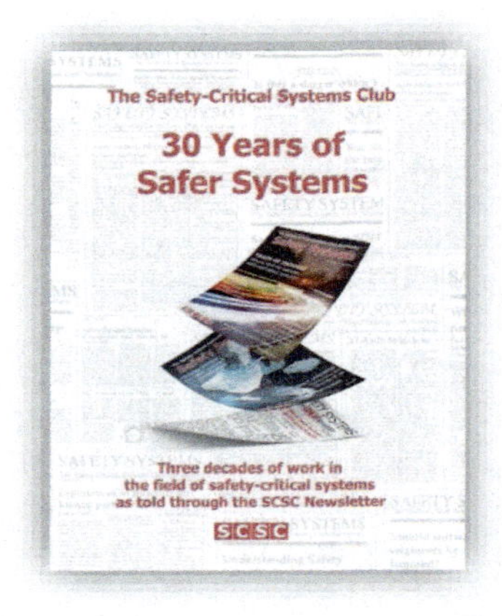

The book groups the articles into themes relevant to safety, with an introduction to the theme and a preface to each article giving major events from the year the article was first published, including accidents, incidents and positive improvements in safety. Themes include: Risk Assessment, ALARP, Artificial Intelligence/Machine Learning, Communication Failures, Safety Culture, 'Black Swan' events, Certification, Product Liability, Safety and Security Integration, Agile Methods, Data driven systems and Safety Cases.

Most of the original authors have provided a short postscript to their article to give extra context and explain progress in the intervening years.

Available for purchase on Amazon
www.amazon.co.uk/Years-Safer-Systems-safety-critical-Newsletter/dp/B09KNCYKDL

The Future of Testing for Safety Critical Systems

THE SAFETY-CRITICAL SYSTEMS CLUB

The Future of Testing for Safety Critical Systems

Thursday 1 December, 2022 - London, UK and blended online

This 1-day seminar will examine how safety-critical systems are tested, some types of testing techniques and the resulting issues. It will also cover emerging technologies, including testing systems including AI and machine learning components. It will cover tools, techniques and limitations of testing methodologies. It will also cover some historical failures of testing and the lessons learnt.

Data Risk Cygnology

Black Swans, Perfect Storms, Dragon Kings and Pudding Lanes – sound like plot elements of an elaborate fantasy novel! However, these are concepts that are relevant to system safety and describe different ways in which events can conspire and lead to catastrophic outcomes. Mike Parsons and Paul Hampton explain four of these concepts from a data perspective.

Black Swan Data

Data that is totally unexpected by those receiving it (i.e. 'out of the blue') and has huge (detrimental) impact, but in retrospect should have been anticipated

In 2013 a C-130 Hercules aircraft landed at Bar Yehuda airport near the Dead Sea [1], a saltwater lake sitting astride Israel and Jordan. The Dead Sea is the lowest place and the airfield lies -1,210 feet below sea level.

The aircraft's navigation system became unresponsive and the constellation of GPS satellites above, mysteriously winked out of existence. As it turned out, the plane's navigation electronics were not designed to operate at altitudes less than 400 feet below mean sea level. In a sense, the plane thought it was underwater!

This is what we call a "Black Swan Data" event; the stream of altitude data going to the navigation systems surprisingly turned significantly negative and could have had a catastrophic result; a situation not considered possible. Yet, with hindsight, and knowledge of earth's geography, we can rationalise this and question why the assumption of positive altitude data was ever thought to be true in the first place.

Might your system be susceptible to Black Swan Data? Presumably not something by its very nature that has been considered; but what can be done to protect against it?

This is an area for further study, but we can offer the following suggestions:

- Be aware of changes of usage, environmental conditions or failures that may produce data that is not expected. The Ariane 501 launch disaster may be thought of in these terms, as the horizontal bias values that caused the exception leading to the complete loss of mission were not anticipated.
- Continually test the established thinking; challenge the oft quoted "that's just how we do things around here".
- What assumptions have we made about the data? Are these really valid and do they continue to be valid in a changing and dynamic world?
- What happens if different inputs (e.g. new sensors or information flows) are added to your system?
- Think outside the box; and we mean, really outside the box. It is the most wild and implausible (compared to current perceptions) ideas that will trip a Black Swan Data event – what could possibly be 'bad data' and how might we handle it?

Perfect Storm Data

Combinations, sets or occurrences of data (or absences of data) that would never have been thought possible to occur together, and when they do have a large and undesirable impact

A Perfect Storm in the risk world is when everything bad happens at the same time, and you did not anticipate that this could happen. These risks get their name from The Perfect Storm, both a 2000 American biographical disaster drama film [2] and a 1997 non-fiction book by Sebastian Junger [3].

In the data world a Perfect Storm is when the data that your system relies upon are all wrong (or missing) at the same time: it can be layers deep. A classic example might be to try and restore deleted data from a backup system, only to find that the backups had not been working properly due to a configuration problem.

So Perfect Storm Data is when multiple data faults or losses conspire to create a situation that has disproportionately catastrophic outcomes.

Another example might be from security: on 10th Dec 2021 a new critical zero-day vulnerability [4] was detected that affected Apache Log4j 2 Java library. It adversely impacted the digital domain and security systems worldwide. The vulnerability, when exploited, permitted remote code execution on the vulnerable server with system-level privileges. The exploit was a combination of both the Java code that contains different logging functions and the settings of a configuration file.

Another example might be the China Airlines A333 at Taipei on 14th June 2020 [5] when all primary computers, reversers and auto brakes failed on touchdown. During landing, flight controls reconfigured from 'normal law' to 'direct law' after all three flight control primary computers (FCPCs) became inoperative. While all aircraft primary control surfaces were still controllable, the deceleration devices including ground spoilers, thrust reversers, and autobrake were lost; the deceleration of the aircraft had to be performed by manual brake.

The following might be considered for management:

- Analyse combinations of data that are important to you: how might one data set being wrong influence another?
- Look at dependencies in your data: are seemingly different sets all derived from a common source?
- Examine as many levels of system and data as you can to establish commonalities and dependencies
- Even if no commonalities are found, consider the unlikely cases of independent data being wrong together, and follow through their consequences

Pudding Lane Data

Data which is unknowingly and unexpectedly critical to the whole operation and if missing or incorrect in some way has a dramatic and negative impact. After the event it may be obvious that it was critical.

"Lord! what sad sight it was by moone-light to see, the whole City almost on fire, that you might see it plain at Woolwich, as if you were by it".

With these words from Samuel Pepys diary in 1666 [6], we see the account of the Fire of London that destroyed 13,200 houses, 87 parish churches, The Royal Exchange, Guildhall and St. Paul's Cathedral, all from a single fire thought to have started in a humble baker's shop in Pudding Lane.

In the same way, unexpectedly critical data can also act as a catalyst to catastrophic outcomes if the dependencies are not identified and adequately mitigated. Often this data is part of an existing system and may be long forgotten about (e.g. a system configuration parameter). It may also be a hard-wired parameter in legacy source code which the maintainers of the code change because they do not understand its original purpose.

In 2015, an Airbus A400M Atlas cargo plane on a test flight crashed at La Rinconada, Spain, less than 5 kilometres from Seville Airport, killing 4 of the 6 crew [7]. Several reports suggested that as many as three of the aircraft's four engines failed during the A400M's departure from Seville.

The suspected cause of the failure was that the torque calibration parameter data had been accidentally wiped on three engines as the engine software was being installed, which would prevent the engine control software from operating properly. In this case, the seemingly innocuous calibration data had a critical role to play in the safety of the entire platform and we introduce the term "Pudding Lane Data" to encompass this form of data. We suggest the following considerations for this type of data:

- Understand and know all your data. The high profile data streams will of course be first and foremost, but do you have an inventory of all the data that the system depends upon and the implications if there is loss of their data properties?
- Configuration and tailoring data may seem static and uninteresting but what are the dependencies on this data? If loss of properties of this data can have catastrophic outcomes, then what processes and mitigations will you put in place to assure those properties are maintained?
- Consider the reporting and monitoring data: if this is incorrect will poor decisions be taken?

Dragon King Data

Data whose effect might have been foreseen, but leads to an unexpected and explosive escalation with major impact ('things get rapidly out of control').

In risk management a Dragon King is defined as an event that is both extremely large in size or impact (a "king") and born of unique origins (a "dragon") relative to other events in the system.

Dragon King events are generated by, or correspond to, mechanisms such as positive feedback, tipping points, bifurcations, and phase transitions that tend to occur in non-linear and complex systems, and serve to amplify events to extreme levels. In the data domain we introduce "Dragon King Data", to represent incorrect or missing data whose consequences unexpectedly escalate into an unpredicted and often catastrophic system event.

An example might be the loss of data relating to Covid-19 test results in the UK due to a row limit in Microsoft Excel. This led to Public Health England losing 15,841 positive test results which in turn meant that 50,000 potentially infectious people were missed by contact tracers and not told to self-isolate [8]. Because of the exponential way that viruses spread, this may have led to many more infections and deaths.

Another, older, example could be the infamous comment given by the weather forecaster Michael Fish about the Great Storm of 1987 [9]:

> *"Earlier on today, apparently, a woman rang the BBC and said she heard there was a hurricane on the way. Well, if you're watching, don't worry, there isn't!"*

These few words may well have led to precautions not being taken across the whole of Southern England, affecting millions of people. In fact the storm was the worst to hit South East England for three centuries, causing record damage and killing 19 people.

Some ways this can be managed are:

- Identify all the receiving systems or users of your data: what are the consequences of errors or losses in the data to those receiving it?
- Analyse the effects of data impact: are any effects subject to non-linear effects? Try to establish the consequence chains and knock-on effects
- Do any of your data potentially impact systems that have large fan-out or spread?
- Create models (or even Digital Twins) of the system in a way that different combinations of data can be tested and simulated to reveal extreme non-linear behaviours

Conclusions

Looking at Data Risks this way can help to tease out some unlikely but severe failures of systems involving data. It is possible that sometimes these risks are already considered in system Hazard and operability study (HAZOP) sessions but discounted as too unlikely to happen. We suggest this approach is a useful and visual 'Tool in the Box' when trying to identify extreme data-related risks of systems. Due to their serious impact we argue that they should be included, assessed and managed alongside more 'normal' risks.

Summary Table

Type	Brief Explanation	Icon
Black Swan Data	Data that is totally unexpected by those receiving it (i.e. 'out of the blue') and has huge (detrimental) impact, but in retrospect should have been anticipated.	
Perfect Storm Data	Combinations, sets or occurrences of data (or sets of absences of data) that would never have been thought possible to occur together, and when they do have a large and undesirable impact.	
Pudding Lane Data	Data that is unknowingly and unexpectedly critical to the whole operation and if missing or incorrect in some way, has a dramatic and negative impact. After the event it may be obvious that it was critical.	
Dragon King Data	Data that might have been foreseen, but leads to an unexpected and explosive escalation with major impact ('things get rapidly out of control').	

References

[1] C-130 Crew Gets a Rude Shock When They Fly Their Plane Below Sea Level, https://www.popularmechanics.com/military/aviation/a26598/c-130-sea-level-dead-sea, accessed August 2022

[2] The Perfect Storm, https://www.imdb.com/title/tt0177971, accessed August 2022

[3] The Perfect Storm: A True Story Of Man Against The Sea, Sebastian Junger, https://www.amazon.co.uk/Perfect-Storm-True-Story-Against/dp/0007230060, accessed August 2022

[4] NCSC Alert: Apache Log4j vulnerabilities, https://www.ncsc.gov.uk/news/apache-log4j-vulnerability, accessed August 2022

[5] A China Airlines Airbus A330 Lost Primary Computers After Landing, https://simpleflying.com/china-airlines-a330-lost-computers, accessed August 2022

[6] The Diary of Samuel Pepys, https://www.pepysdiary.com/diary/1666/09, accessed August 2022

[7] 2015 Seville Airbus A400M crash, https://en.wikipedia.org/wiki/2015_Seville_Airbus_A400M_crash, accessed August 2022

[8] PHE statement on delayed reporting of COVID-19 cases, https://www.gov.uk/government/news/phe-statement-on-delayed-reporting-of-covid-19-cases, accessed August 2022

[9] Quotes of famous people, https://quotepark.com/quotes/1829009-michael-fish-earlier-on-today-apparently-a-woman-rung-the-bbc, accessed August 2022

Mike Parsons, SCSC DSIWG Chair

Mike is the SCSC Director and Events Coordinator. He also leads the SCSC Service Assurance, Data Safety Initiative and the new Systems Approach to Safety of the Environment Working Groups. He is currently a researcher at the University of York working on the Assuring Autonomy International Programme (AAIP). He has been in the business of safety since 1989.

Paul Hampton, SCSC Newsletter Editor

Paul is a Chartered Engineer with over 30 years' experience in IT. He has spent 15 of those designing and developing enterprise systems in many diverse sectors and the remainder involved in system safety in a variety of capacities including: safety engineering, safety management, independent auditing and corporate governance and assurance.

Risk Matrices and Pitfalls

Risk is a concept we implicitly understand as human beings, and we discuss and evaluate it in many situations on a day-by-day basis. But what is the actual definition of risk? And what is considered an acceptable/tolerable risk? Does this concept change depending on the context, or has it an absolute meaning? How do we represent the risk? Lucia Capogna and Stephen Bull explore these questions and provide advice on some of the pitfalls that may be encountered when categorising and treating risk.

Risk definition – a reminder

In this article we will try to answer the questions around risk and present one of the possible representations and method for risk ranking: the risk matrix. Examples, types of matrices, their calibration and the main pitfalls are also discussed.

The International Electrotechnical Vocabulary (IEV) defines risk as "combination of the probability of occurrence of harm and the severity of that harm".

This definition has been adopted by many standards, including EN 61508-4:2010 and ISO 26262-1:2018.

A slightly deviation from this definition of risk is documented in the EN50126-1:2017, where "risk, <for railway RAMS>," is defined as "combination of expected frequency of loss and the expected degree of severity of that loss".

Thus, "risk" can be seen as a two-dimensional concept, where the two dimensions are the probability or frequency of occurrence of an event and the severity of the consequence of that event.

What is a risk matrix?

A Risk Matrix is a common method used for risk ranking or categorisation.

A Risk matrix is defined by Cox (2008) [1] as "a table that has several categories of "probability," "likelihood," or "frequency" for its rows (or columns) and several categories of "severity," "impact," or "consequences" for its columns (or rows, respectively)".

There are three types of risk matrices:

- qualitative risk matrix: the items on the axes of the matrix are defined with qualitative or descriptive terms (not numerical)
- quantitative risk matrix: the items on the axes of the matrix are defined in quantitative or measurable terms (numerical)
- semi-quantitative matrix: it is a combination of the qualitative and quantitative matrices. One of the axes can be defined qualitatively and the other one is expressed quantitatively (or one or both axes could be defined both qualitatively and quantitatively).

This is an example of a qualitative risk matrix (although it could be made quantitative by defining ranges for the descriptive frequency and severity labels).

Frequency of occurrence of an accident (caused by a hazard)	Risk Acceptance Categories			
Frequent	Undesirable	Intolerable	Intolerable	Intolerable
Probable	Tolerable	Undesirable	Intolerable	Intolerable
Occasional	Tolerable	Undesirable	Undesirable	Intolerable
Rare	Negligible	Tolerable	Undesirable	Undesirable
Improbable	Negligible	Negligible	Tolerable	Undesirable
Highly improbable	Negligible	Negligible	Negligible	Tolerable
	Insignificant	Marginal	Critical	Catastrophic
	Severity of an accident (caused by a hazard)			

For reference, CENELEC, EN 50126-1:2017 Annex C [5] gives an explanation of risk matrices (although no guidance on their calibration).

Where are risk matrices used?

Risk matrices, independently from the type, are commonly used to rank/categorise risks in a hazard log to prioritise effort and/or to single out high risks for further attention.

Later in the lifecycle they are used to accept risks in a hazard log.

Risk matrices are also used in many other fields than safety to determine the acceptability of other types of risks, e.g. they are also used for determining the acceptability of the operational, security, reputation and environment risks.

Example risk matrices

Examples of risk matrices can be found in many standards or codes of practice applicable to different industries or sectors, but also in guidance created for specific projects (usually where the project is complex or high profile).

With regards to the railway industry, an example risk matrix can be found in the EN 50126-1:2017 standard. This has been also reproduced in the Network Rail (NR) guidance on CSM-RA. Project-specific examples include the recently opened Crossrail in London; many other examples or guidelines can be found.

Other examples can be found in the Defence Industry:

- the Regulatory Article 1210 (RA 1210) issued by the Military Aviation Authority and specific for the ownership and management of operating risks, and
- the MIL-STD-882E issued by the US Department of Defence. This standard identifies the Department of Defence (DoD) approach for identifying hazards and assessing and mitigating associated risks encountered in the development, test, production, use, and disposal of defence systems.

Pitfalls in the definition and the use of risk matrices

When looking into the definition of risk matrices, we should consider that there are multiple exposed populations in different roles (e.g. workers and passengers) and the (perceived) risk appetite may be different for each. Thus, the risk matrices specific for each population are likely to be different. (For example, the UK HSE document Reducing Risks, Protecting People (R2P2) [3] assigns different targets for workers and members of the public.)

The judgement around the frequency, the severity or the risk appetite is often subjective and dependent on the experience of the people involved in their definition/calibration, in hazard workshops and in hazard review meetings.

In addition, the risk matrices definition should not be a standalone exercise but should consider the big picture (including applicable law or regulations and possible contradictions), otherwise across the same project/programme or domain the adopted risk matrices may be inconsistent. This does not mean that risk matrices for different operational aspects should be identical, but that it is very important they are compatible.

This should be also considered when different organisations are involved in the same programme. A supplier/subcontractor may have their own risk matrix which is not (necessarily) aligned to the programme / operator's matrix. The specific project scope (or portion of the project scope) should be considered in the risk matrix definition, as the overall risk matrix may be not adequate for the scope of the analysis or the sub-system. This adds additional burden to the entity in charge of the integration and responsible to "import" the contribution from a different matrix with possible transformation consequence.

This practice attracted criticism from the Haddon-Cave report into the Nimrod XV230 crash [2], which said "There are a myriad of different Hazard Risk Matrices used to determine risk categorisations. This is confusing, potentially dangerous, and makes it more difficult to compare risks across platforms", and we share that view that multiple risk matrices can cause confusion. However, for the reasons presented in this paper it is essential to calibrate matrices to the individual application. It is also essential that the reasons for different matrices are clearly justified and that care is taken to ensure that matrices used within the same project or programme reach consistent conclusions on the level of risk.

When using a risk matrix, it is also fundamental to consider the level of detail at which individual events are assessed. If the risk matrix is calibrated (see below) to consider hazards at the boundary of a system then separate assessment of the individual causes of the hazard (each representing a smaller portion of the overall risk) may lead to the false conclusion that the risk is acceptable. This needs to be taken into account:

- when deriving a risk matrix for a particular programme/project and
- when applying the risk matrix to individual hazards.

When using the risk matrices to rank risks, there are a few other areas that can lead to misuse or misinterpretation of the risk matrices too. For example, the link between tolerable and ALARP. There is the misconception that a tolerable risk is automatically ALARP. However, they are two different concepts: it is still necessary (and legally required in the UK) to demonstrate that all practicable measures have been taken to reduce the risk (ALARP). Hence, a risk matrix cannot be used on its own to declare that a risk is ALARP!

"There are a myriad of different Hazard Risk Matrices used to determine risk categorisations. This is confusing, potentially dangerous, and makes it more difficult to compare risks across platforms." (Haddon-Cave)

Another important pitfall in the use of risk matrices is the understanding and interpretation of the risk matrix axes.

The frequency of occurrence can be understood as "frequency of the Event (often the hazard) that might lead to the consequence" or the "estimated frequency of the Consequence". Many people do not understand the difference and mix them up in the same analysis. Instead, the consequence may be understood as the "worst case consequence" or the "most likely reasonably foreseeable consequence". It is very important to not mix up these concepts in the same analysis and understand the difference in their interpretation.

Sometimes standards and codes of practice may help in solving this. For example, in the railway sector a clear definition of the axes is documented in the EN 50126-1:2017, that states that the axes should be interpreted as the frequency and consequence of the accident arising from the hazard.

Another aspect to consider is the "additionality": if the scope of the programme is increased during its execution, then an amount of extra risk is likely to be added: depending on the nature of this risk, the risk matrix may need recalibrating and existing risks may need re-evaluating.

For further information, RSSB, Taking Safe Decisions, 2019 [4] provides useful principles on how to evaluate safety of a system.

Calibration

It is possible to calibrate a risk matrix from "first principles", taking into account that there are lots of issues and shortcomings. Often this calibration is necessary to ensure that the matrix is appropriate for the project.

When the risk is derived from a generic (overall) target, the calibration of the risk matrix must consider the scope of the analysis and how much of the overall risk can reasonably be assigned to the system under consideration.

The calibration should also consider the level of the risk assessment, i.e. if the scope is the assessment of individual hazards, then only a proportion of the risk budget (i.e. the tolerable level of risk for the overall system) is available per hazard and the risk matrix needs to be scaled accordingly.

For further reading on calibration, GE/GN8642 Issue 2 (withdrawn, but available from RSSB) (Appendix C) provides some guidance on calibrating a risk matrix [6].

Conclusions

The use of the risk matrices is considered an "easy" approach and for this reason they are very often overused or abused.

In reality, they are not really so easy to use, as they require to be calibrated to their particular application and there are many possible pitfalls in their use or definition.

Risk matrices can be a useful tool to quickly assess risks if properly customised / calibrated, however there are contexts where the use of risk matrices is not appropriate and alternative techniques are available.

Any use of a risk matrix should be accompanied by expert judgement as to the suitability and relevance of the outputs; it must also consider the context and the possible misuses and misinterpretations. Perhaps any risk matrices should come with a comprehensive set of assumptions and warnings!

References

[1] Cox Jr., "What's Wrong with Risk Matrices? Risk Analysis 28", L.A. 2008

[2] Charles Haddon-Cave QC, "The Nimrod Review - An independent review into the broader issues surrounding the loss of the RAF Nimrod MR2 aircraft XV230 in Afghanistan in 2006", Oct. 2009

[3] HSE, Reducing Risks, Protecting People (R2P2), 2001: HSE's general approach to risk management, including the "carrot diagram" and overall targets for acceptable risk

[4] RSSB, Taking Safe Decisions, 2019: gives principles about how to evaluate safety of a system

[5] CENELEC, EN 50126-1:2017 Annex C: gives an explanation of risk matrices (no guidance on their calibration)

[6] GE/GN8642 Issue 2 (withdrawn, but available from RSSB) (Appendix C) provides some guidance on calibrating a risk matrix

Lucia Capogna, Principal Engineer, Ebeni Limited

Lucia has over 15 years' experience in software, safety and cyber security with background in multiple industries including Railway, Defence, Oil & Gas and Renewable Energy. Lucia also represents the UK in several CENELEC and IEC standardisation groups for software, cyber security and safety standards.

Stephen Bull, Principal Safety Engineer, Ebeni Limited

Stephen has over 25 years' experience in the development and assurance of safety critical systems in a variety of industries, including rail, defence and air traffic control. He is a Chartered Engineer and a Fellow of the Safety and Reliability Society.

Managing Unexpected Risks Event Report

The Seminar "Managing Unexpected Risks: Handling Rare and Severe Events Now and in the Future", took place in person at the BCS offices in London and online on 26th May 2022. Paul Hampton and Martin Atkins provide the highlights of the event.

The event opened with Mike Parsons, SCSC Managing Director, introducing the concepts of 'Black Swans'. He said that increasingly complex systems and new technologies like autonomous vehicles are presenting new opportunities for the occurrence of Black Swan events.

Managing Novel Incidents

The first presentation was by John Holmes from NATS, the UK air traffic service provider, on the topic of Managing Novel Incidents. John talked about an incident associated with a well-established system.

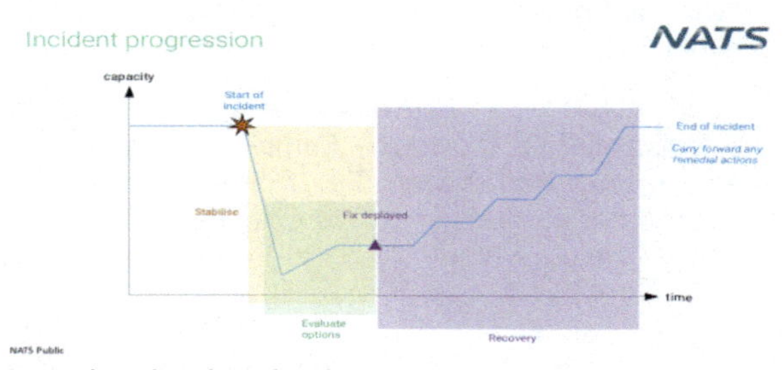

Traffic was low due to an overnight storm and there was an intermittent fault with a radar display system in the morning resulting in delays. Intermittent low-level packet loss was being experienced on a network switch, but it was decided not to replace it, because of the storm. However, a switch on the redundant network then also failed, resulting in track data blocks – data associated with the trails displayed on Air Traffic Control (ATC) displays – being intermittently lost.

To deal with the situation, NATS implemented traffic control methods, by reducing the number of aircraft taking off. The fault was diagnosed and a fix was completed in around 4 hours.

There is a mature hierarchical incident management system at NATS and incidents are escalated based on the level of impact. This is well-exercised through training. However, analysis of the event found that:

- There was loss of a single point of truth between ATC and Engineering
- Focus was on the controller positions but less on other affected systems connected to the same network
- There was inconsistent use of language and communication between different stakeholders
- There was insufficient recording of the decision-making process to provide evidence that the correct actions had been taken

Although the team arrived at what proved to be the correct course of action, these good decisions were not an inevitable product of the process.

Following an investigation, NATS developed a more prescriptive process for evaluating options for risks, benefit and trade-offs and a set structure for recording decision rationale. This was based around a structured series of questions, but was not adopted in the end, as it was felt to be cumbersome and risked stifling the flexible and reactive nature of the more informal problem solving approach.

It was acknowledged that there is heavy reliance on highly trained, skilled, adaptable and empowered people and NATS decided to create a better training programme for staff to ensure there is a clear framework for necessary discussions, retaining flexibility and pace.

Black Swans - Decisions under Uncertainty

David Slater from Cardiff University presented on the importance of instinct, experience, and time for thinking in emergencies.

"Normal" emergencies are usually well-handled as the system is well understood but 'black swans' are different as we are dealing with the 'Unknown Unknowns' as famously expressed by Donald Rumsfold.

David introduced a range of complexity categorisations from Obvious or Simple (the known), Complicated (the knowable), Complex (the unknowable), Chaotic (incoherent), Disorder (not determined). He introduced a 'Cygnology' for these systems and gave examples: for Complicated emergencies, he gave examples of Bhopal and Piper Alpha and for Chaotic emergencies, he gave Black Swan events, such as Pandemics.

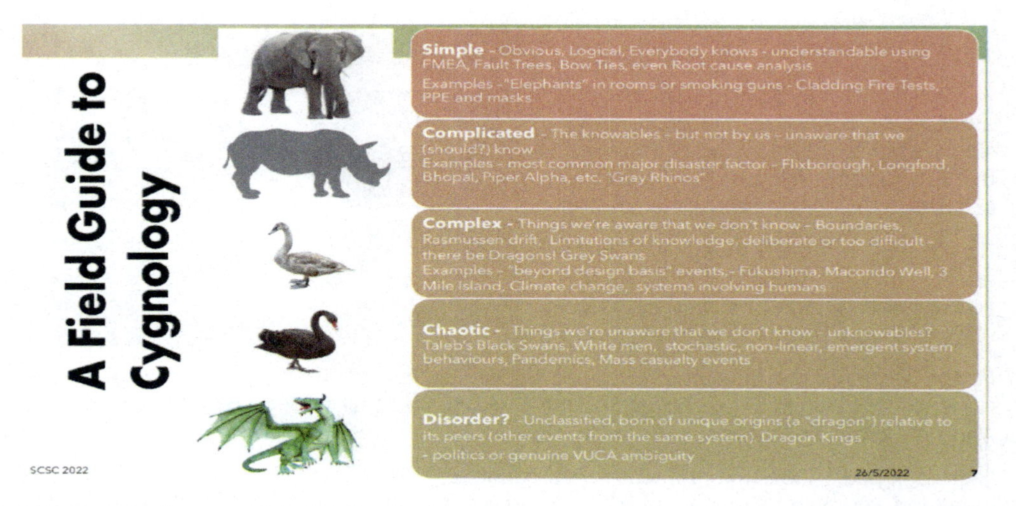

The Cynefin framework helps the decision-making process but David noted that there can be drift between quadrants (Rasmussen Drift).

David gave further examples of Simple, Complicated and Complex accidents and gave examples of real Black Swans in the Chaotic category such as 9/11, Covid-19 and the shelling of Ukrainian Nuclear Power stations.

David then went on to discuss what we can do about managing these types of events. There is acceptance that Chaos does happen, we live in a complex environment, but we have evolved to survive in these environments. Planning and training are seen as important but is 'robotic' Plan Do Check Act (PDCA) methodologies sufficient? David discussed other models such as John Boyd's OODA loop (Observe, Orient, Decide and Act). Klein also had a model called Recognition Primed Response. But the real question is whether there is time to think logically or read the manual during a crisis.

David also referred to Kahneman's thinking 'fast and slow' with system 1 being automatic and little or no effort and no sense of voluntary control. System 2 allocates attention to the effortful mental activity.

He referred to Libet's experiment and the complexity of the brain.

David noted how people respond differently to emergency situations and referred to the Fukushima Daini (No 2) plant which survived a similar set of problems to the Daiichi (No 1) plant that experienced the well-publicised meltdown. The No 2 plant was saved due to innovative human ingenuity to help route power after the flooding by the tsunami. David thought that AI systems can help in the simple or more complicated systems, but their inflexibility could limit their utility in more complex situations.

David concluded that the instinct, experience and time for thinking in emergencies is critical. It's important to have more than just classroom-based training, there should be empowerment of those responsible, leadership not micromanagement, team leading and building resilience.

The Buncefield Explosion

Martin Atkins from MCA Ltd discussed the events surrounding the Buncefield Explosion. He defined a Black Swan as an event that is a surprise, has a major impact and in hindsight, is rationalised as predictable.

He stressed however, that Black Swans are very much 'in the eye of the beholder' and relative to role an affected party plays in an incident.

The Fuji Building

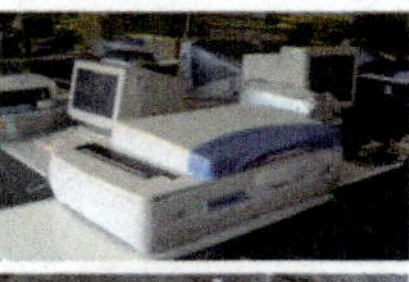

Martin discussed the Buncefield Site in 2005, an Oil Storage Terminal near Hemel Hempstead, the 5th largest oil storage depot in the UK with a capacity of 60,000,000 gallons and filling ~400 tankers a day.

Each tank had 3 indicators if it was full: a gauge, another (which was not operating at the time) and a backup full/empty switch that was intended to be an independent safety mechanism.

For tank number 912, only the backup was operational and the gauge was known to stick occasionally. A lever selected whether the backup was looking for full or empty and was supposed to be locked in the full position. However, the padlock to protect this was never fitted and there is some doubt of which mode was selected.

There were also design and construction issues: the tank construction caused overflowing fuel to "bounce" and increase evaporation, and the walls ("bunds") between tanks could not contain a spill due to gaps and bad gaskets.

Early in the Sunday morning of the incident, the tank was being remotely filled and started to overflow, with about 250,000l of fuel making a pool 2m deep. There were then several explosions with the largest one being a "fuel-air" explosion, which was felt 28 miles away. This led to a fire that burned for 3 days.

This in itself was not a Black Swan event, being a conventional failure of safety mechanisms. However, when considered from the perspective of organisations situated in buildings adjacent to the site, this was clearly a Black Swan event. One building in particular, the Northgate building, contained a datacentre which contained the Schengen Information System, the Police National Computer's hot stand-by backup system and the Addenbrookes Hospital Admission and Discharge system. Further afield, other consequences included the closure of a motorway, fuel rationing of flights at Heathrow, closures of schools and public buildings, and pollution from the fire and the fire-fighting chemicals.

Martin then discussed an issue in January 2021 where data was accidentally deleted from the Police National Computer; it took ~15 weeks to recover the records, which could not have been helped by the hot stand-by backup system, destroyed during Buncefield event, never having been replaced. Martin argued that although the event itself was a normal predicted event, it became a Black Swan event for many people outside, such as businesses, a hospital, the Police, the environment etc.

Martin concluded that there will always be Black Swan events, but w
Black Swan depends on who you are.

Martin suggested that to protect against these events requires
pendent on the nature of the disaster, such as Disaster Rec
resilience. Organisations need to take a wider view, not just w
including their suppliers, the physical surroundings and how failures
may affect them.

Hunting Black Swans in Power Grids Related to EV Ch
Satellite Navigation Timing Signals

Andrew Larkins from Sygensys Ltd discussed Black Swans in the context of Power grids.

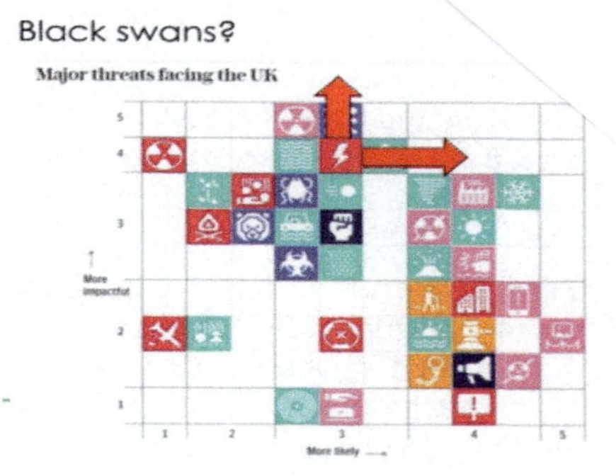

Failures of power grids typically arise from severe weather, failures and supply imbalance and these can have significant impact as there is an extreme societal dependency on power. For example, outages in 2021 in Texas led to many businesses being left without power for several days and there were hundreds of fatalities associated with the event.

The dependency on power is reflected in the National Risk Register with impact at the high end and medium likelihood.

Andrew said that the introduction of electric vehicles (EV) will increase the load on the grid and represents a rapid change that will present a significant challenge to grid operators – by 2035 charging is predicted to be over 40% of the total load at certain times of the day.

A key question is whether the system will still work when faced with these external challenges, and so, this might be a good hunting ground for Black Swans. Andrew found that there were no detailed investigations about the impact of the short-term stability of EV charging after a fault, to prevent a "domino" effect occurring.

Andrew said to help hunt potential Black Swans you need time, skill and freedom to think around the issues.

EVs, by their nature, are high demand and users will want to charge on low tariffs in a smart optimised way, and there is the potential to manage this using IOT technologies. This means that the operator can shift from varying the supply to match the demand, to controlling the demand itself, but this introduces new failure modes.

Andrew suggested some of types of teams and techniques required to explore Black Swans, such as cross-functional teams, mix of generalist and specialists and trying to get people to think in different ways.

ults of a number of workshops in this area resulted in 6 ways in which EV chargers esent a risk to grid security. For example:

- Step: Too many chargers switching on or off at the same moment
- Ramp: Too many chargers switching on or off within a few minutes
- Oscillations: a group of chargers switching on and off repeatedly
- Degraded stability: increases risk of post-fault collapse
- Demand control: deficiencies are eroded
- Restoration: Erratic behaviour after restart hindering process of restoration.

Andrew asked whether it can be classed as Black Swan if it is predicted before it happened. His focus is on how to fix these future issues but noted there is, for various reasons, resistance from industry to focus on this area at the moment.

Andrew concluded with a discussion on Satellite navigation signals, which are subject to Jamming and Spoofing despite the high dependency that modern society has on them, and so are potential future sources of Black Swan incidents.

Grey Swans

Bill Blackburn from Process Renewal Group concluded the afternoon session with a discussion on Grey Swans. Bill noted that things do happen, such as financial collapses and security breaches that affect large volumes of people.

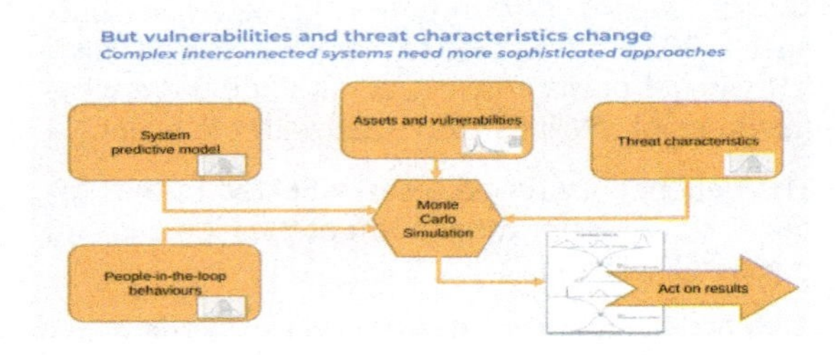

Bill noted that there are a large number of regulators but noted that it is impossible to regulate away risk and regulators are quite fragmented. Often companies align their organisations around the regulator requirements, compliance is more 'tick box' and regulators struggle to keep abreast of change.

Bill then showed the national risk register and noted that there are several risks that are clearly safety-critical but others that might be classed as safety-related such as Cyber attacks, serious and organised crime and animal diseases.

Bill discussed what items were not in the national risk register such as the Stephen Hawking warning around AI being the 'best or worst thing' for humanity, which then brought in the question around ethics.

Other areas are Quantum computing, which might bring benefits, such as improving the accuracy of inertial GPS systems (eg. using quantum accelerometers and gyroscopes), but might have downsides, such as undermining current encryption techniques. Bill asked how best to gain this foresight and noted that some government organisations like the Department for Business, Energy & Industrial Strategy (BEIS) are doing horizon scanning and commissioning others like the Royal Society to glimpse into future technologies, such as computer/neural interfaces.

Bill said that Quantum Computing, AI and Ethics are hot topics for lawyers who are exploring these areas as potential future areas of litigation.

Bill said the UK Financial Conduct Authority (FCA) are building resilience into financial services after various crashes and discussed some of the challenges the FCA see, such as, technical innovation (AI, blockchain, crypto assets), changing behaviour, keeping pace, challenging environments and system complexity.

Assessing threats in the real world is expensive and can be dangerous and Bill noted some commonly used approaches – techniques (such as those in SCSC Guidance documents), system modelling, crisis-management and simulations.

Vulnerabilities also evolve, for example, behaviourally with people moving to higher risk areas, the demand for greater mobility and overuse of drug/antibiotics. New business models and technology adoption and supply chains can often mask vulnerabilities and can have exposure to cyber threats.

Bill concluded that the Royal United Services Institute's (RUSI) viewpoint of the National Risk Register is that half of the risks are missing. Having an entry on the register does not mean that we have methods to manage the risk.

Safety and Resilience of Designs of the Future

Wendy Owen from Bangor University discussed resilience and the work she has been involved in with the Institute of Asset Management (IAM) working group producing SSG-32 Contingency Planning & Resilience Analysis.

Wendy started by explaining the terminology around resilience, resilience analysis and contingency planning. She discussed the resilience analysis process and some points around resilient organisations being typically forward thinking.

Resilience At A Glance

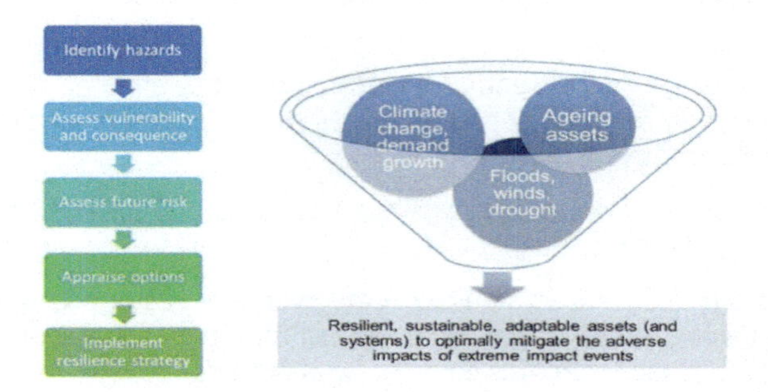

She gave examples of the typical processes where resilience analysis is undertaken and contingency planning activities are developed.

- Prevention: technique used to prevent the issue from occurring
- Response: what is done to respond to an issue once it has occurred
- Recovery: how to return the business back to normal operations and importantly want to learn lessons and prevent reoccurrences

Wendy provided several case studies that were in SSG-32 such as the Pacific Gas & Electric San Bruno (2010) rebuilding after a gas explosion.

In this study, after the incident, rather than abandoning the site, they rearchitected the site with automation, replacement of pipeline, testing and enhanced telemetry and set their goal to be the safest and most reliable company in the industry.

Wendy also noted that there are standards for Business Continuity Management (BCM) such as those used by Network Rail. Some BCMs can use questionnaires to help structure assessments.

Space Weather Risks

The final presentation was pre-recorded by David Southwood and Jonathan Eastwood from Imperial College and introduced in the meeting by Bill Blackburn. David introduced Jonathan as an expert in Space and Atmospheric Physics at Imperial College London.

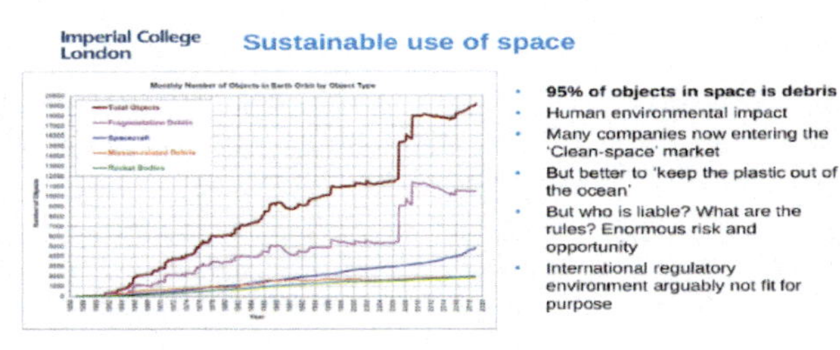

Jonathan explained what Space Weather meant and he noted it was on the national risk register. He said auroras are the most recognisable visible part of a much larger system resulting from solar activity. Solar flares are common but there are also Coronal Mass Ejections (CME) from time to time that take about 1 or 2 days to reach the Earth but can have significant interaction with the Earth's Magnetosphere.

Jonathan then discussed the worst that could happen from a severe space weather event, typically those that might occur every 100 years. There is real concern that it can disrupt the supply of electricity. Other impacts would be PNT (Position Navigation and Timing) signals from GPS. In space, it can affect the services provided by satellites, and the satellites themselves.

Jonathan gave an example, of a weather event in 1989 that affected supplies in Quebec, North America. However, this was not as severe as that thought to have occurred in 1859, known as the 'Carrington' event.

In another example, in February 2022 SpaceX launched 40 Starlink satellites, but lost most of these because of a small geomagnetic storm.

Space weather is hard to assess as many extreme events have not occurred in the modern technological era. Each event is unique and could last for a period in the order of two weeks. Jonathan asked for feedback from this community about this topic.

Jonathan made some observations:

- The economic impact could be huge based on studies
- Many organisations are not aware of their reliance on space infrastructure and how their products may be exposed to space weather
- There is much more predicted use of satellites in space, such as ubiquitous internet access, and this will bring new challenges, such as managing the 95% of objects now in space, which are debris

Jonathan concluded by asking what a reasonable response is to space weather risk?

The Future of Coding for Safety-Critical Systems

The Future of Coding for Safety-Critical Systems Seminar aimed to provide insights into the future of coding for safety systems: what languages, techniques and methodologies are likely to be used, what coding standards and guidelines may be applicable and how verification and analysis might be integrated with the process of developing code. Dewi Daniels provides the highlights from this fascinating event.

The SCSC held a seminar on "The Future of Coding for Safety-Critical Systems" on Thursday, 9th June 2022. This was a blended event, held in-person at the Wellcome Collection, Euston Road, London and online via Zoom. This was the second SCSC seminar to be held at the Wellcome Collection, which was a lovely venue that was a very short walk from Euston Square underground station.

The C Language... You Wouldn't Start from Here!

The first talk was by Andrew Banks from the Motor Industry Software Reliability Association (MISRA) on "the C language... you wouldn't start from here!" Andrew said he was sorry that Rod Chapman hadn't been able to attend, since he and Rod like to taunt each other about the relative merits of C and Ada!

Andrew started with a brief history of C, followed by the history of MISRA C. He described MISRA C and showed how MISRA C checking would have found Apple's SSL bug [1].

Safety For Software-Intensive Systems

The second talk was by Paul Sherwood from CodeThink on "safety for software-intensive systems". Paul started by describing the history of CodeThink and their safety journey in automotive, including Artificial Intelligence (AI) and infotainment.

Paul presented a proposal as to how to define trustable software. He said that most software is not certified, or even certifiable, and he was sceptical of partitioning claims. He cited an example where an untrusted JavaScript client could crash the hardware.

He presented a solution using Nancy Leveson's Systems Theoretic Process Analysis (STPA) to model the functionality required to support safety applications and examine what could go wrong. His use of STPA is supplemented by fault injection testing to show that the mitigation is effective.

What About Rust for Critical Systems?

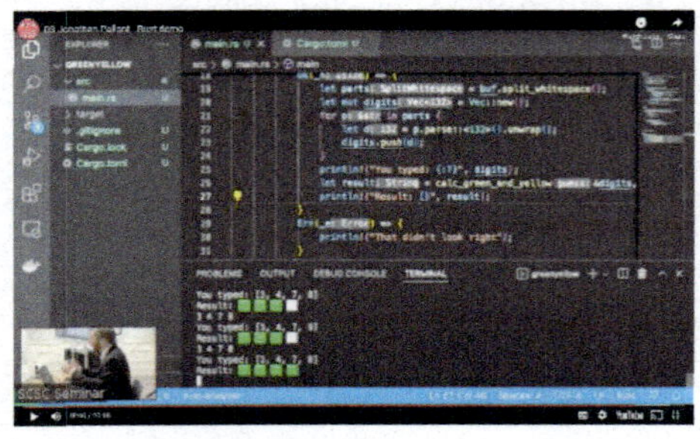

The third talk was by Jonathan Pallant from Ferrous Systems on "what about Rust for critical systems?" The talk was originally to be presented by Rod Chapman of Protean Code, but Rod was feeling unwell. Rod is well-known for his work on the SPARK programming language, and he has been using and studying Embedded Rust for about 4 months.

Jonathan felt it would be useful to present a third-party (i.e., Rod's) view on the Rust programming language. John is very well-qualified to speak about Rust, having written two chapters of the Embedded Rust Book [2]. Jonathan gave an introduction to the Rust language, including an interactive demonstration where he wrote a Rust program to implement a number-guessing game, and shared Rod's thoughts on the language, which were that Rust is a huge leap forward compared to C but that, compared to SPARK, there are some rather obvious omissions.

Auto Code and Auto Verify – Is That All?

The fourth talk was by Nick Tudor from D-RisQ on "auto code and auto verify – is that all?" Nick said there had been a great reaction to his paper in the inaugural SCSC Journal [3].

Nick said that people like to write code, but they're not so interested in software verification. A lot of software projects go over-budget. We can't just test and say that's good enough.

Nick advocated removing the opportunity for error introduction by using suitable languages and tools, and using mathematical techniques to replace testing, while hiding the mathematics from the user. He described a set of tools developed by D-RisQ, including Kapture: a requirements management tool, Modelworks: a design verification tool, CLawZ: a source code verification tool and FEVER: an executable object code verification tool.

Coding With Meaningless Symbols

The final talk was by Les Hatton, emeritus professor of Forensic Software Engineering at Kingston University, on "coding with meaningless symbols". Les no longer works in safety-critical software – he says he can't stand it! He now works as a geneticist.

Les said there is good software out there, but we don't have the knowledge to build software as a scientific discipline. We can recognise good and bad code, but we can't explain why. While there are systems out there that are obviously very good, we need to spend more effort on finding out what makes them good.

Les showed that there are emergent and apparently inevitable properties of all discrete systems, from the length of words in books, the distribution of wealth, the length of proteins and the length of subprograms in software programs. Whatever programming language you choose and whatever your program does, your system as a whole is overwhelmingly likely to share the same emergent properties, safety-critical or not.

Since his talk, Les has published an excellent book [5] on this topic, which is available as a paperback or for the Kindle. I have read the book and highly recommend it.

Where To Find Out More

More details of the event, including videos of the presentations and copies of the presentation slides, can be found on the SCSC website at Seminar: The Future of Coding for Safety-Critical Systems (scsc.uk) [4]

References

[1] Apple's SSL/TLS bug (22 Feb 2014), https://www.imperialviolet.org/2014/02/22/applebug.html, retrieved 28 July 2022.

[2] The Embedded Rust book https://docs.rust-embedded.org/book/, retrieved 28 July 2022.

[3] Daniels & Tudor, Software Reliability and the Misuse of Statistics, SCSS Journal, Vol. 1, No. 1, Software Reliability and the Misuse of Statistics (scsc.uk), retrieved 28 June 2022.

[4] Seminar: The Future of Coding for Safety Critical Systems https://scsc.uk/e912

[5] Exposing Nature's Bias: The Hidden Clockwork Behind Society, Life and the Universe, Les Hatton and Greg Warr, Bluespear Publishing, 10 June 2022, ISBN 978-1908422040

Dewi Daniels, Software Safety Limited

Dewi is a highly experienced software engineer specialising in the development and verification of safety-critical and other kinds of high-integrity software. He has worked on programmes ranging from civil airliners such as Airbus A380 and Boeing 787 to personal air vehicles, air taxis and unmanned air systems.

Safety Futures Initiative Update

Zoe Garstang, lead for the Safety Futures Initiative (SFI), provides an update on the progress made by the SFI including details of future events, and she asks for volunteers for the SFI's new mentoring scheme.

Progress Update

Work is underway to provide SFI members with a series of events and activities. This includes a mentoring scheme and regular events, such as further Get To Know You sessions and an upcoming lecture competition (Spring 2023).

The lecture competition will see individuals deliver a short presentation on a safety-related topic of their choice, followed by a short Q&A. Keep an eye on the SFI website (www.scsc.uk/gf) for updates and to find out when entries open.

Mentoring Scheme – Volunteers Needed!

We are establishing a mentoring scheme for SFI members and members of the SCSC community to build on knowledge sharing and career development. The scheme will be one-to-one mentoring, which is flexible to suit the requirements and schedules of those involved. The focus of the mentoring sessions can be anything from technical areas through to career development.

If you would like to register your interest as a mentor or mentee please contact Zoe Garstang (zoe.garstang@scsc.uk).

Upon registering your interest, you will be provided with a short questionnaire in order to match you with the best possible mentor/mentee that aligns with your experience and aspirations.

> **"We are establishing a mentoring scheme ... to build on knowledge sharing and career development."**

Membership

Membership for the SFI is free for the first year, so I would encourage anyone who would like to get involved to sign-up (please see www.scsc.uk/membership).

SFI members get access to all SFI events and activities, as well as discounted fees at SCSC Events.

Further Information

If you would like further information about any aspect of the SFI, please do get in touch with Zoe Garstang (zoe.garstang@scsc.uk).

Zoe Garstang, Airworthiness Engineer and SFI Lead

Zoe is an Airworthiness Engineer at BAE Systems. She previously undertook an Advanced Engineering Apprenticeship with the company before joining the Airworthiness & Product Integrity Team.

How Do I Get Into Safety?

It is sometimes difficult to plan a career: so many things have to align together for each step on the ladder to take place, whether it is the correct training, organisation, job or assignment, or simply just being in the right place at the right time. We are sometimes asked: "how do I get into safety?"

The answer isn't straightforward. In fact, it is doubly difficult to plan a career in safety engineering, assurance, or consultancy as a solid background in the underlying technologies (such as software, architectures, or databases) and sector knowledge (eg. aviation, nuclear or rail), plus the right opportunities all have to be present

Safety is often a second career, taken on by engineers or consultants who have already got several years of experience doing other things. Also, there is the issue that not many safety staff are required for most jobs: often there is only one safety engineer on smaller projects; if that role is already taken, there is little chance of a junior gaining relevant experience.

With this situation in mind, we continue our series of articles where members of the SCSC Steering Group share their experiences of 'getting into safety'. Some of the routes taken are definitely not linear!

Please read and compare with your own 'safety story'. Of course, these experiences are based on events some time ago, and the situation has definitely improved as the industry has matured. For example, there are now courses (at post-graduate level) on safety-critical systems such as those at the University of York, there are competency frameworks for safety roles, and the SCSC has its own "Safety Futures Initiative" programme with the aim of developing young and early-career staff so that they can take on full safety roles.

The messages that come out of these experiences however, are that sometimes you do just have to be in the right place at the right time, and with the right underlying characteristics. All safety roles require the ability to be able to assess risk, to understand some difficult technical arguments, to follow (and create) workable processes, rules and regulations, to know the standards and guidelines relevant to the job in hand, to be able to communicate well, work in a team with colleagues, and very importantly, to be assertive and take a strong position when needed.

Roger Rivett

I worked for 10 years developing software for engine management systems. In 1990, the department head, on reading an article about applying BS 9000 to software development, decided that the department should work within a software quality management system (QMS). Of his staff at the time, he considered me to be best person to make this happen. Despite my initial reservations, I accepted the new role of software quality manager with the brief to introduce a software quality management system with external certification.

In the same time frame, the MISRA group was formed under the DTI SafeIT programme with the brief to produce an automotive response to the recently issued draft of what became IEC 61508. This led to the publication in 1994 of the Development Guidelines for Vehicle Based Software. I was my company's representative in the MISRA group, this gave me the opportunity to explore the subject in more detail and I found the topic both intellectually stimulating and of great importance to the wellbeing of my company's customers. Consequently, I started incorporating the functional safety recommendations from the MISRA guidelines into the requirements of the software QMS.

When the software QMS was considered to be firmly established, after successfully completing its 3-yearly recertification, I handed over day to day responsibility for it to a member of my staff and focussed on functional safety, as by now this had become my main area of interest. This involved performing hazard analysis and risk assessment on all the electronic control units fitted to the vehicle at that time and developing associated procedures for inclusion in the QMS. Further knowledge and understanding came from regular attendance at conferences, especially the yearly Safety Systems Symposium run by the SCSC, involvement in further DTI and European projects and continuing involvement with MISRA. The latter resulting in the publication in 2007 of the Guidelines for safety analysis of vehicle based programmable systems. During this time, I became the company's Technical Specialist for functional safety and in 2005 I became a founder member of the ISO working group that developed ISO 26262, first published in 2011.

Jane Fenn

Whilst working as a Systems Engineer for BAE Systems, it was noted that I also had a flare for Project Management, so when a safety-critical system displayed some worrying behaviour during rig testing, they asked me to put together a multi-disciplinary team to undertake a full safety assessment and agree a modification programme with the supplier.

During that time, I met people who became mentors and friends, who drew me further and further into the dark arts of Safety Engineering. I became entirely and shamelessly hooked. When a replacement system was planned, I was determined that we would engineer 'in' safety from the outset. Defence Standard 00-55 issue 2 had only recently been issued and we followed that closely, bringing a level of rigour that was unusual at that time. I started to undertake modules of the Safety Critical Systems Engineering MSc course at the University of York, to ensure I was applying latest and best practice.

Needless to say, once that programme was finished, I jumped at the chance when I was offered the opportunity to manage BAE Systems' research centre, the 'Dependable Computing Systems Centre' at York, and in conjunction with Newcastle University; it was my dream job! That allowed me to be involved with and influence leading edge Safety Engineering research.

Subsequent roles included managing industrial research teams, across defence industry, developing modular safety cases, and later, software architectures, as well as managing the MOD-funded pan-defence industry Software Systems Engineering Initiative. My current role is 'Lead Consultant Engineer – Engineering Safety' where I define the Safety Engineering policy, competency and training requirements as well as some of the processes used by safety practitioners within BAE Systems Air Sector.

I dabble in the occasional safety consultancy activity with other Business Units of BAE Systems and support several of the SCSC Working Groups: Assurance Case, Safety Culture, Security-Informed Safety and Safety of Autonomous Systems. And those people who led me astray all those years ago.... some of them are on the SCSC Steering Group today....

Dewi Daniels

I graduated with a degree in Computing Science from Imperial College, London in 1981. My first job after graduation was with Logica. I spent the 1980s developing software that was not safety-critical but was mission-critical. For example, I designed a software product that allowed applications running on VAX minicomputers to interface to airline reservations networks. I was privileged to demonstrate the product to Ken Olsen (founder of DEC) at DECworld '87. I also worked on a secure trading system for the Bank of England.

I first got involved with safety-critical software when I went to work for Program Validation Limited (PVL) in 1989. There, I was one of the developers of the SPARK Examiner. SPARK is still widely used, though the software we wrote has long since been replaced by open-source software developed by AdaCore. I also did program proof for several Rolls-Royce jet engines, including the RB211-535. This was one of the first industrial applications of program proof, as far as I know, and was very labour-intensive, though also very enjoyable (interactive program proof felt a lot like playing Tetris).

In the mid-1990s, I worked for Lloyds Register. I spent 6 months in Atlanta working on the software safety case for the then-new Lockheed C-130J (which is to be retired from RAF service next year!). I also spent 6 months in Paris writing the outline safety case for the European Geostationary Navigation Overlay Service (EGNOS), a precursor to Galileo. In the late 1990s, I worked for Praxis Critical Systems. It wasn't a very happy time for me; Praxis wanted me to do system safety engineering, whereas I was much more interested in software development.

In 2004, I went to work for Silver Software, which was a great company and a great bunch of people. A turning point for my career was when I joined the committee that was writing DO-178C. I greatly enjoyed the experience and made many friends. In 2010, I was headhunted to form Verocel Limited, the UK subsidiary of a US company. I grew Verocel Limited from a one-man start-up to a company with a turnover of over £2 million. I recruited a great team of software engineers. I also started a company called Aeronautique Associates in 2012 to offer DO-178C training. When I left Verocel in 2015, I decided to do some consulting while I decided what to do next. I've found that I really enjoy being an independent consultant. I particularly enjoy the variety of work. I've worked on some very interesting projects. I led a team that developed an innovative flight control system for an optionally manned rotorcraft. I helped to develop an unmanned air system (UAS) autopilot. I've audited several companies against DO-178C and DO-254. I've always been interested in software processes, so I enjoy seeing how different teams approach software and hardware development. I'm currently helping Vertical Aerospace build an air taxi. My consultancy work has enabled me to do unpaid work that I also enjoy. I'm currently a member of the RTCA/EUROCAE Forum on Aeronautical Software and of EUROCAE WG-117. I'm one of the UK representatives on the IEC 61508 committee. I'm a member of the SCSC Steering Group.

SCSC is my main means of interacting with, and learning from, like-minded professionals. What I particularly value about the SCSC is the mix of industrial and academic members.

Sean White

I have only ever worked as a Safety Engineer so know nothing else – hang-on a minute, as a Safety Engineer you need to know a lot about a lot of things – that's what's kept me enthralled, challenged and motivated over a number of decades.

I started in the role, although I didn't know it at the time, as a fresh faced electronics Graduate, working in the Advanced Engineering Department at Westland Helicopters. Days when integrated modular avionics, fly by wire and active control technologies were in their infancy and safety standards were at best interim drafts. The standards matured and so did my understanding and skills and I still get a buzz seeing a Merlin hovering in the sky.

A move to BAe and fast jets gave me the opportunity to work on Hawk, Tornado, Typhon and best of all Harrier – it must be the hovering thing. A short secondment to Boeing to develop a safety case for the Harrier flap controller and a longer one to Flight Test to support Hawk AJT development were highlights.

I then jumped ship or is that plane to join the NHS, working within a team to establish a safety management system for health technologies. Challenging, frustrating but also very rewarding. I'm now writing national & international standards and training clinicians and technologists in the principles of safety management. Still actively involved in research, investigating how to assure AI technologies for health and medical purposes.

60 Seconds with ... Les Hatton

Les Hatton Ph.D. is a mathematician notable for his work on failures and vulnerabilities in software-controlled systems. Originally an award-winning geophysicist, he switched careers in the early 1990s to study software and systems failure.

He has published widely and his theoretical and experimental work on software systems failure can be found in "Nature", "IEEE Transactions on Software Engineering", and numerous other journals. His 1995 book "Safer C" helped promote the use of safer language subsets in commercial embedded control systems.

In 2012, he took early retirement to try to understand recurring patterns he was observing in the many millions of lines of code he had then studied. In 2014 he proved that all software systems asymptote to the same length and alphabet distribution. In 2015, with the biochemist Greg Warr, he extended this to the distribution of known proteins and has since demonstrated similar behaviour in software, biology, music, literature and many other systems. Their findings have been recently published in the 2022 book: "Exposing Nature's Bias: the hidden clockwork behind society, life and the universe".

He is an emeritus Professor of Forensic Software Engineering at Kingston University, London and spends his spare time doing computational biology in his shed!

What first attracted you to working in the field of software reliability?

I was an earth scientist for the first 20 years of my career. In meteorology and later in seismology I found glaring examples of software failures frequently derailing the science. In 1992, I changed careers to study software reliability to try and find answers.

What aspect of your career are you most proud of?

I don't look back so taking early retirement in 2012 to work full-time on a problem in information theory with an old college friend, a biochemist, which eventually led to the discovery of a new Law of Nature – the Fundamental Law of Inequality. This has been hidden in plain sight for 5,000 years, but manifests itself in literature, language, wealth distribution, software and notably, genomics. The book just came out after some 15 years of work. Curiously, although unintentional, this mostly resolved my original questions about software reliability, but its implications for evolutionary biology are particularly exciting.

What advice would you give to yourself age 12?

Practice the guitar more.

What future changes would you like to see in the field of System Safety?

The primacy of engineering professionalism and judgement over management, media and indeed politics.

What's your most favourite quote or motto?

Albert Einstein. When he heard that a book titled 100 Authors against Einstein was published in Germany, he said:

"If I were wrong, then one would have been enough."

If you could learn to do anything, what would it be?

Fly.

"Taking early retirement in 2012 to work full-time on a problem in information theory with an old college friend ... eventually led to the discovery of a new Law of Nature – the Fundamental Law of Inequality"

If you could be any fictional character, who would you choose?

None of them really. Fictional characters are defined by somebody else. I'm far happier with my own path. If I could play any fictional character, it would probably be "the man with no name", although I'm not very good with horses.

What's the best piece of advice you've ever been given?

In spite of his unfailing belief in the opportunities of education, my dad's career advice was:

"It's not what you know, it's who you know".

The younger idealistic me disagreed, but boy was I wrong.

Which work of art or fiction best sums up your experiences with Covid-19?

It would have to be a work of fiction – the government's Covid-19 figures. I volunteered to work in a Covid modelling group through the pandemic and the data quality was truly awful – some of the worst I've ever had to work with.

Recent Publications

Guidance on the Safety Assurance of autonomous systems in Complex Environments.
http://www.york.ac.uk/assuring-autonomy/guidance/sace/

Revealing a newly discovered fundamental law of inequality.
http://www.amazon.co.uk/Exposing-Natures-Bias-Clockwork-Universe/dp/1908422041

Safety-Critical Systems eJournal vol.1 no.2
Summer Issue 2022
scsc.uk/scsc-176

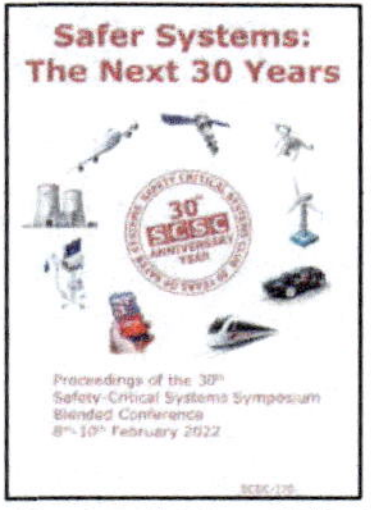

Proceedings of the 30th Safety-Critical Systems Symposium.
scsc.uk/scsc-170

Guidance on the management of data safety risks.
scsc.uk/scsc-127G

Three decades of work in the field of safety-critical systems as told through the SCSC Newsletter.
scsc.uk/scsc-169

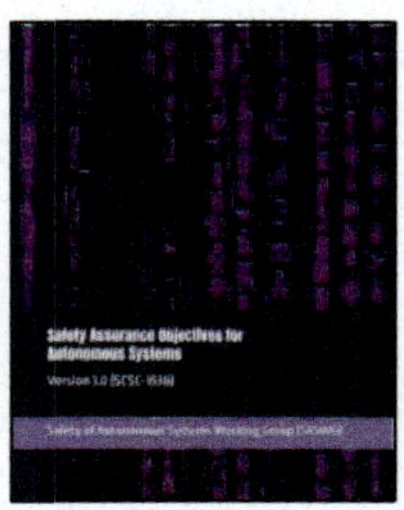

Guidance on assuring autonomous systems.
scsc.uk/scsc-153B

Guidance on assuring safety-related service.
scsc.uk/scsc-156B

Assurance Case Guidance covering Challenges, Common Issues and Good Practice.
scsc.uk/scsc-159

Book Review

Dewi Daniels reviews the new book, "Exposing Nature's Bias: The Hidden Clockwork behind Society, Life and the Universe", by Les Hatton and Greg Warr.

Les Hatton, who is well-known to members of the SCSC, has co-authored a new book [1] that presents a ground-breaking new theory on the emergent behaviour of all discrete systems.

Les presented some of the early research that led to this discovery at the 25th Safety-Critical Systems Symposium [2]. More recently, Les gave a very interesting presentation on the topic at an SCSC seminar [3]. Les will be giving the after-dinner speech at SSS'23.

Benoit Mandelbrot claimed that power law distributions are ubiquitous. Les and Greg have been able to explain why, for the first time. They've called their discovery the Fundamental Law of Inequality.

They show how inequality is the overwhelmingly most likely outcome of all discrete systems. Their theory is corroborated by analysis of large amounts of data, which has only been made possible in the last few years by the availability of powerful personal computers and the publication of open databases.

We think of evolution as an inexorable advance towards organisms of increasing complexity and that humans are the pinnacle of evolution. The Fundamental Law of Inequality shows that nearly all evolution occurs within simple organisms. An alien observer could quite reasonably conclude that bacteria are the dominant form of life on Earth.

The book also shows that inequality of wealth has always been with us and is likely to always remain with us. The Scandinavian countries have low levels of wealth inequality and are among the happiest in the world. Have the Scandinavians found the secret of happiness, or are they simply the countries that happen to be in the long tail of the distribution?

This is one of the most thought-provoking books I've read in the past few years.

References

[1] Exposing Nature's Bias: The Hidden Clockwork behind Society, Life and the Universe, Les Hatton and Greg Warr, Bluespear Publishing, 10 June 2022, ISBN 978-1908422040 https://www.amazon.co.uk/Exposing-Natures-Bias-Clockwork-Universe/dp/1908422041

[2] Balancing safety with rampant software feature-itis, Les Hatton, Twenty-fifth Safety-Critical Systems Symposium, 9 February 2017, Bristol UK, https://scsc.uk/r135/24:1

[3] Coding with meaningless symbols, Les Hatton, SCSC Seminar on the Future of Coding for Safety-Critical Systems, London, 9 June 2022, Seminar: The Future of Coding for Safety-Critical Systems https://scsc.uk/e912prog

Connect

The Newsletter and eJournal

Do you have a topic you'd like to share with the systems safety community? Perhaps an interesting area of research or project work you've been involved in, some new developments you'd like to share, or perhaps you would simply like to express your views and opinions of current issues and events. There are now two publishing vehicles for content – shorter, more informal content, can be published in the Newsletter with longer, more technical peer-reviewed material more suitable for the eJournal. If you are interested in submitting content, then get in touch with Paul Hampton for Newsletter articles: paul.hampton@scsc.uk or John Spriggs for eJournal papers: john.spriggs@scsc.uk

The SCSC Website

Visit the Club's website thescsc.org for more details of the Safety-Critical Systems Club including past newsletters, details of how to get involved in working groups and joining information for the various forthcoming events.

Facebook

Follow the Safety-Critical Systems Club on its very own Facebook page.

www.facebook.com/SafetyClubUK

Twitter

Follow the Safety-Critical Systems Club's Twitter feed for brief updates on the club and events: @SafetyClubUK

LinkedIn

You can find the club on LinkedIn. Search for the Safety-Critical Systems Club or use the following link:

www.linkedin.com/groups/3752227

Advertising

Do you have a product, service or event you would like to advertise in the Newsletter? The SCSC Newsletter can reach out to over 1,000 members involved in Systems Safety and so is the perfect medium for engaging with the community. For prices and further details, please get in touch with the Newsletter Editor.

Safety Systems Crossword

Welcome to the first Safety Systems Crossword prize competition where you can test your knowledge of the world of System Safety!

Send you completed grids to crossword@scsc.uk by 31st Dec 2022. The first correct answer drawn from a hat at SSS'23 will win a £50 Amazon Voucher!

Across

7. Abbreviation for a system designed to prevent aircraft collision (4)
9. A device for keeping an aircraft on a set course without the intervention of the pilot (9)
10. Type of attack that targets computer information systems, computer networks, infrastructures, or personal computer devices (5)
11. Type of process automation that mimics back-office tasks of human workers (7)
13. Abbreviation for a technology with applications such as natural language processing (2)
15. Type of network for representing knowledge about an uncertain domain (8)
19. Name given to the event associated with the most intense geomagnetic storm in recorded history (10)
20. Farmer from Kansas who founded an aircraft manufacturing company (6)
23. A visual display device used in the railway industry to convey information to drivers (6)
24. Author of a report that led to the development of a national police intelligence system (7)
25. Type of stochastic process conditional only on the present state of the system (6)
27. Actual colour of a flight data recorder (6)

Down

1. Abbreviation for a system-theoretic model of accidents (5)
2. A type of tube for measuring fluid flow velocity (5)
3. Abbreviation for a type of top-to-bottom failure analysis technique (3)
4. Location of a significant nuclear accident precipitated by a powerful earthquake (9)
5. Name of the vehicle where cold weather contributed to its loss (10)
6. Low Earth Orbit satellite constellation company that survived Chapter 11 bankruptcy (6)
8. Site of a nuclear power station deriving its name from "fort on the mound" (8)
12. Adjust the aerodynamic forces on the control surfaces so that an aircraft maintains the set attitude without any control input (4)
14. A feature that makes the state of two mechanisms or functions mutually dependent (9)
16. A type of chart that has a graphic representation of a sea area and adjacent coastal regions (8)
17. Shutdown a nuclear reactor in an emergency (5)
18. Type of machine learning algorithm that learns patterns from untagged data (12)
21. A device which detects or measures a physical property and records, indicates, or otherwise responds to it (6)
22. Vehicle safety feature that reduces the severity rather than likelihood of an accident (6)
23. Nickname of the pilot involved in a famous emergency water landing (5)
26. Abbreviation for a type of graphite-moderated nuclear power reactor (4)

SCSC Working Groups

The Safety-Critical Systems Club is committed to supporting the activities of working groups for areas of special interest to club members. The purpose of these groups is to share industry best practice, establish suitable work and research programmes, develop industry guidance documents and influence the development of standards.

Assurance Cases

The Assurance Cases Working Group (ACWG) has been established to provide guidance on all aspects of assurance cases including construction, review and maintenance. The ACWG will:

- Be broader than safety, and will address interaction and conflict between related topics
- Address aspects such as proportionality, rationale behind the guidance, focus on risk, confidence and conformance
- Consider the role of the counter-argument and evidence and the treatment of potential bias in arguments

In Aug 2021, the group published v1.0 of the Assurance Case Guidance: scsc.uk/scsc-159

One of the working group's activities is the maintenance of the Goal Structuring Notation (GSN) Community standard.

See scsc.uk/gsn for further details.

In May 2021, the group published v3.0 of the standard: scsc.uk/scsc-141C

Lead Phil Williams phil.williams@scsc.uk

SCSC Working Groups

Security Informed Safety

The Security Informed Safety Working Group (SISWG) aims to capture cross-domain best practice to help engineers find the 'wood through the trees' with all the different security standards, their implication and integration with safety design principles to aid the design and protection of secure safety-critical systems and systems with a safety implication.

The working group aims to produce clear and current guidance on methods to design and protect safety-related and safety-critical systems in a way that reflects prevailing and emerging best practice.

The guidance will allow safety, security and other stakeholders to navigate the different security standards, understand their applicability and their integration with safety principles, and ultimately aid the design and protection of secure safety-related and safety-critical systems.

Lead **Stephen Bull** stephen.bull@scsc.uk

Data Safety Initiative

Data in safety-related systems is not sufficiently addressed in current safety management practices and standards.

It is acknowledged that data has been a contributing factor in several incidents and accidents to date, including events related to the handling of Covid-19 data. There are clear business and societal benefits, in terms of reduced harm, reduced commercial liabilities and improved business efficiencies, in investigating and addressing outstanding challenges related to safety of data.

The Data Safety Initiative Working Group (DSIWG) aims to have clear guidance on how data (as distinct from the software and hardware) should be managed in a safety-related context, which will reflect emerging best practice.

An update to the guidance (v3.4) was published in Jan 2022: scsc.uk/scsc-127G

Lead **Mike Parsons** mike.parsons@scsc.uk

SCSC Working Groups

Safety of Autonomous Systems

The specific safety challenges of autonomous systems and the technologies that enable autonomy are not adequately addressed by current safety management practices and standards.

It is clear that autonomous systems can introduce many new paths to accidents, and that autonomous system technologies may not be practical to analyse adequately using accepted current practice. Whilst there are differences in detail, and standards, between domains many of the underlying challenges appear similar and it is likely that common approaches to core problems will prove possible.

The Safety of Autonomous Systems Working Group (SASWG) aims to produce clear guidance on how autonomous systems and autonomy technologies should be managed in a safety-related context, in a way that reflects emerging best practice.

The group published v3 of its guidance Safety Assurance Objectives for Autonomous Systems, in Jan 2022 scsc.uk/scsc-153B

Lead **Philippa Ryan** pmrc@adelard.com

Multi- and Manycore Safety

It is becoming harder and harder to source single-core devices and there is a growing need for increased processing capability with a smaller physical footprint in all applications. Devices with multiple cores can perform many processes at once, meaning it is difficult to establish (with sufficient evidence) whether or not these processes can be relied upon for safety-related purposes.

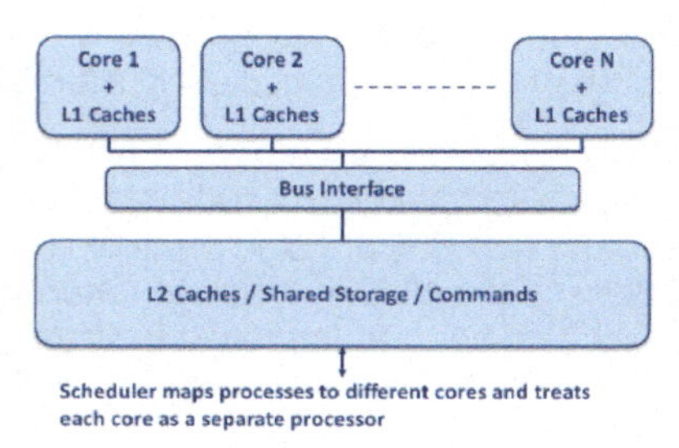

Parallel processes need to access the same shared resources, including memory, cache and external interfaces, so they may contend for the same resources. Resource contention is a source of interference which can prevent or disrupt completion of the processes, meaning it is difficult to know with a defined uncertainty the maximum time each process will take to complete (Worst Case Execution Time, WCET) or whether the data stored in shared memory has been altered by other processes.

The Multi- and Manycore Safety Working Group (MCWG) has been established to explore the future ways of assuring the safety of multi- and manycore implementations.

Lead **Lee Jacques** Lee.Jacques@leonardocompany.com

SCSC Working Groups

Ontology

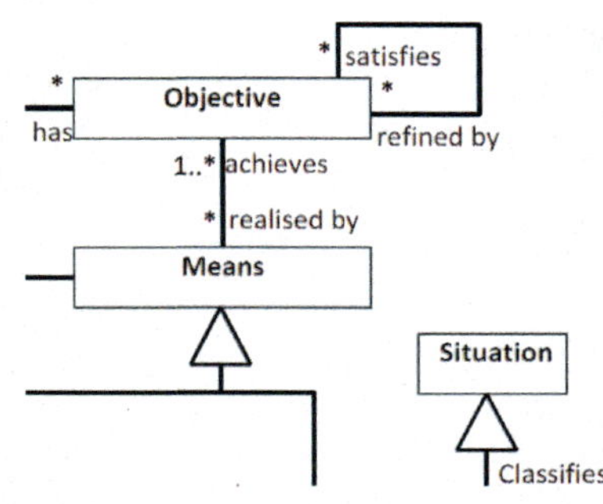

The Ontology Working Group (OWG) develops ontologies that will form the basis of SCSC guidance, as well as having wider industrial and academic applications.

The OWG is currently working on the definition of an ontology of risk for application in guidance for risk-based decision making – notably safety and security – and for which ISO 31000 Risk Management principles are to be applied.

The Data Safety Working Group (DSIWG) developed the core aspects of the Risk Ontology, which has been migrated to this working group. The Risk Ontology will form the upper ontology to the Data Safety Ontology that the DSIWG will continue to develop.

Lead **Dave Banham** ontology@scsc.uk

Covid-19

The Covid-19 Working Group is involved with discussion, analysis and assistance related to the Coronavirus. The group meets remotely to see what a systems and assurance view of the situation brings.

The group has compiled an extensive range of Covid-19 related material and made this available on the working group's website pages along with ongoing developments in the thoughts and ideas of the group.

Members are all experienced engineers, used to making reasoned arguments about safety. The aim is to apply the groups considerable technical expertise to the problem and find and assure appropriate solutions.

Lead **Peter Ladkin** ladkin@causalis.com

SCSC Working Groups

Service Assurance

Risks presented by safety-related services are rarely explicitly recognised or addressed in current safety management practices, guidelines and standards. It is likely that service (as distinct from system) failures have led to safety incidents and accidents, but this has not always been recognised. The Service Assurance Working Group (SAWG) has been set up to produce clear and practical guidance on how services should be managed in a safety-related context, to reflect emerging best practice.

The group published v3.0 of the guidance in Jan 2022: scsc.uk/scsc-156B

Lead Mike Parsons mike.parsons@scsc.uk

SCSC Safety Culture

The Safety Culture Working Group (SCWG) has been established to provide guidance on creating and maintaining an effective safety culture. The group seeks to improve safety culture in safety-critical organisations focussed on product and functional safety, by sharing examples and latest approaches collated from real-life case studies.

Meetings provide an opportunity to discuss any particular aspects attendees are interested in taking forward, and to help set future directions for the group.

Lead Michael Wright michael.wright@greenstreet.co.uk

Systems Approach to Safety of the Environment

The Systems Approach to Safety of the Environment Working Group (SASEWG) is a new group intending to apply Systems Safety practices to systems that are embedded within the natural environment, while focussing on that environment.

The group aims to produce clear guidance on how engineered systems should be developed and managed throughout their entire lifecycle so as to preserve, protect and enhance the environment.

The group held its inaugural meeting on 17th June 2022 with its 3rd meeting planned for 19th October 2022. Please get in touch with the working group lead if you would like to join, or find out more about this group.

Lead Mike Parsons mike.parsons@scsc.uk

SCSC Membership

The SCSC provides a range of services to the System Safety community including seminars, tutorials, leadership events, specialist topic working groups, the annual symposium and a comprehensive body of publications. Membership brings many valuable benefits such as free access to online events, the SCSC Newsletter and access to presentations and other resources from events.

Individual Membership

To become an individual member of the SCSC please register on the SCSC website using the icon at the top right of any page and select "Register". Complete and save your account registration and then verify your email address. Once registered and logged in click the link "why not join the SCSC..." inviting you to become a member at the top right of the page or select "Pay membership" from the icon.

Individual membership can be paid online using a credit/debit card through our secure payment partner Realex Global Payments or contact Alex King for other payment methods. For student or retired member rates please contact Alex King to get your account status changed.

Corporate Membership

Your company contact with the SCSC should arrange the membership and any renewals for your organisation. To join as a member covered by a corporate membership, register as per the instructions for an individual member and then contact Alex King to confirm your affiliation.

Renewing Membership

You should be notified by email when your membership is almost expired or shortly after it has expired. These notifications will contain a link to the online renewal page or you will be able to renew when logging onto the website through the 'click to renew' link.

Membership Fees

The following fees are applicable for new and renewing members:

- 1 year Individual Membership: £125
- 2 year Membership: 20% discount: £200
- 3 year Membership: 33% discount: £250 (3 years for the price of 2)
- 1 year SFI Membership: FREE for first year, £35 for years 2 & 3
- 1 year Membership, retired member rate: £35
- For Corporate Membership discounts contact Alex King.

A one-month Publication Pass is also available for £15. This allows access to all SCSC website publications in a particular calendar month.

Contact Alex King using office@scsc.uk

The SCSC Steering Group

Tom Anderson
Honorary member

Stephen Bull
stephen.bull@scsc.uk

Dai Davis
Honorary member

Zoe Garstang
zoe.garstang@scsc.uk

Louise Harney
louise.harney@scsc.uk

Brian Jepson
brian.jepson@scsc.uk

Graham Jolliffe
Honorary member

Alex King
alex.king@scsc.uk

Wendy Owen
wendy.owen@scsc.uk

Felix Redmill
Honorary member

John Spriggs
john.spriggs@scsc.uk

Phil Williams
phil.williams@scsc.uk

Robin Bloomfield
Honorary member

Dewi Daniels
dewi.daniels@scsc.uk

Jane Fenn
jane.fenn@scsc.uk

Paul Hampton
paul.hampton@scsc.uk

James Inge
james.inge@scsc.uk

Nikita Johnson
nikita.johnson@scsc.uk

Tim Kelly
Honorary member

Mark Nicholson
mark.nicholson@scsc.uk

Mike Parsons
mike.parsons@scsc.uk

Roger Rivett
roger.rivett@scsc.uk

Emma Taylor
Honorary member

Sean White
sean.white@scsc.uk

Club Positions

The current and previous (marked in italics) holders of club positions are as follows:

Managing Director

Mike Parsons 2019-

Tim Kelly 2016-2019

Tom Anderson 1991-2016

Programme & Events Coordinator

Mike Parsons 2014-

Chris Dale 2008-2014

Felix Redmill 1991-2008

Newsletter Editor

Paul Hampton 2019-

Katrina Attwood 2016-2019

Felix Redmill 1991-2016

eJournal Editor

John Spriggs 2021-

Website Editor

Brian Jepson 2004-

Steering Group Chair

Roger Rivett 2019-

Graham Jolliffe 2014-2019

Brian Jepson 2007-2014

Bob Malcolm 1991-2007

Manager

Alex King 2019-

Honorary Solicitor

Dai Davis 2022-

University of York Coordinator

Mark Nicholson 2019-

Administrator

Alex King 2016-

Joan Atkinson 1991-2016

Safety Futures Initiative Lead

Zoe Garstang 2019-

Nikita Johnson 2019-2021

Calendar

October '22

M	T	W	T	F	S	S
					1	2
3	4	5	6	7	8	9
10	11	12	13	14	15	16
17	18	19	20	21	22	23
24	25	26	27	28	29	30
31						

November '22

M	T	W	T	F	S	S
	1	2	3	4	5	6
7	8	9	10	11	12	13
14	15	16	17	18	19	20
21	22	23	24	25	26	27
28	29	30				

December '22

M	T	W	T	F	S	S
			1	2	3	4
5	6	7	8	9	10	11
12	13	14	15	16	17	18
19	20	21	22	23	24	25
26	27	28	29	30	31	

January '23

M	T	W	T	F	S	S
						1
2	3	4	5	6	7	8
9	10	11	12	13	14	15
16	17	18	19	20	21	22
23	24	25	26	27	28	29
30	31					

February '23

M	T	W	T	F	S	S
		1	2	3	4	5
6	7	8	9	10	11	12
13	14	15	16	17	18	19
20	21	22	23	24	25	26
27	28					

March '23

M	T	W	T	F	S	S
		1	2	3	4	5
6	7	8	9	10	11	12
13	14	15	16	17	18	19
20	21	22	23	24	25	26
27	28	29	30	31		

April '23

M	T	W	T	F	S	S
					1	2
3	4	5	6	7	8	9
10	11	12	13	14	15	16
17	18	19	20	21	22	23
24	25	26	27	28	29	30

May '23

M	T	W	T	F	S	S
1	2	3	4	5	6	7
8	9	10	11	12	13	14
15	16	17	18	19	20	21
22	23	24	25	26	27	28
29	30	31				

June '23

M	T	W	T	F	S	S
			1	2	3	4
5	6	7	8	9	10	11
12	13	14	15	16	17	18
19	20	21	22	23	24	25
26	27	28	29	30		

July '23

M	T	W	T	F	S	S
					1	2
3	4	5	6	7	8	9
10	11	12	13	14	15	16
17	18	19	20	21	22	23
24	25	26	27	28	29	30
31						

August '23

M	T	W	T	F	S	S
	1	2	3	4	5	6
7	8	9	10	11	12	13
14	15	16	17	18	19	20
21	22	23	24	25	26	27
28	29	30	31			

September '23

M	T	W	T	F	S	S
				1	2	3
4	5	6	7	8	9	10
11	12	13	14	15	16	17
18	19	20	21	22	23	24
25	26	27	28	29	30	

Events Diary

11 Oct 2022
Conference

High Integrity Software (HIS)

Marriott Hotel City Centre, Bristol

www.his-conference.co.uk

12-14 Oct 2022
Conference

RISK/SAFE 2022: 13th Conference on Risk Analysis, Hazard Mitigation and Safety and Security Engineering

Rome, Italy

www.wessex.ac.uk/conferences/2022/risk-safe-2022

18 Oct 2022
SCSC Working Group

Assurance Case Working Group (ACWG) #17

Online

scsc.uk/e958

19 Oct 2022
SCSC Working Group

Systems Approach to Safety of the Environment (SASEWG) #3

Zoom, 16:00-17:30

scsc.uk/e964

20 Oct 2022
SCSC Working Group

ACWG: Assurance Case Paper TG

Online

scsc.uk/e960

3 Nov 2022
SCSC Working Group

ACWG: GSN SWG

Online

scsc.uk/e959

22-23 Nov 2022
Conference

10th Scandinavian Conference on System and Software Safety

Gothenberg, Sweden

safety.addalot.se/2022

28 Nov 2022
SCSC Working Group

SCWG: SCSC Safety Culture Working Group (SCWG) Meeting

Location – TBA

scsc.uk/e922

1 Dec 2022
SCSC Seminar

The Future of Testing for Safety Critical Systems

London, UK and blended online

scsc.uk/e966

7-9 Feb 2023
SCSC Symposium

Safety Critical Systems Symposium SSS'23

York, UK and online

scsc.uk/e898

6 Apr 2023
SCSC Seminar

Safety of Autonomy in Complex Environments

London, UK and blended online

scsc.uk/e890

Partner Event

The High Integrity Software conference is back! #HISConf2022

www.his-conference.co.uk

After two years of virtual events, HIS 2022 will be making its physical comeback on Tuesday 11th October 2022 at a new venue: the Bristol Marriott Hotel City Centre for what promises to be a great event.

Modern-day national infrastructure programs are often comprised of critical software-intensive systems. A key differentiator with these projects is the scale of the engineering endeavour coupled with the need to remain operational for the long term – often over several decades.

This year, the High Integrity Software conference will consider this issue from a technical perspective. More specifically, the aim is to understand how early decisions around cyber-physical system architectures and adopting effective software development lifecycles and verification techniques will later impact defect rates and maintainability. However, the challenge for such large-scale and long-term systems is multi-faceted. Therefore, the conference will also consider broader issues within the software development ecosystem, such as sustainable supply chains and talent streams.

To see the full programme and book your place - https://www.his-conference.co.uk/

All delegate tickets are £220.00 (excl VAT) per person

If you have any questions about the conference, the exhibitor package or further sponsorship opportunities, contact info@his-conference.co.uk.

Printed in Great Britain
by Amazon

17626121R00115